D0475989

Cattle of the World

CATTLE of the World

John B. Friend, B.Sc.(Agric)

Illustrations by Denis Bishop

BLANDFORD PRESS

Poole Dorset

First published 1978

Link House, West Street
Poole, Dorset BH15 1LL

ISBN 0 7137 0856 5

Set in Great Britain by
Dorchester Typesetting Co. Ltd.

Printed by
Sackville Press Billericay Ltd.,
Billericay, Essex

Bound by
Robert Hartnoll Ltd.
Bodmin

CONTENTS

Author's Acknowledgements 6
Illustrator's Acknowledgements 7
Preface 9
1 The Cattle Family 10
2 Domestication of Cattle 11
3 Cattle Migrations 15
Cow Parts 24
4 Cattle Breeds 26
Breed Societies and Herd Books 38
Milk Road Transport 50
Beef 74
Cattle as Draught Animals 86
Hides to Leather 110
Milk Production 120
Milking Techniques. 134
Cattle Rail Transport 146
Milk Rail Transport 147
5 Milk and Milk Products. 158
Well Known Milk Products 160
Butter and Cheesemaking 168
6 Modern Breeding Techniques 170
'The Fifth Quarter' 176
7 Shows and Sales 185
Bibliography 194
Index 195

AUTHOR'S ACKNOWLEDGEMENTS

Many people have helped me during the writing of this book and I would like to express my gratitude to them all. None deserve more thanks than my wife Mary, who has encouraged me and taken over many of my home responsibilities whilst I was writing. Our two young children, Duncan and Fiona, also deserve a special mention for allowing me to write undisturbed. The manuscript was typed by my mother and father, and I am very grateful to them for this valuable help.

My thanks must go to Denis Bishop for involving me in this project in the first instance. The excellent illustrations executed by him and his son Simon inspired me to try and achieve the same standard in my writing. I also owe a great deal to my good friend and colleague Wendy Symonds whose constructive criticism has made this book far more understandable than it might otherwise have been.

Other personnel at the Milk Marketing Board must also be thanked; particularly Pauline Collins and her staff in the MMB Library who made me welcome in spite of my frequent demands on their time. Without the facilities of the MMB Head Office, the task of writing this book would have proved very difficult. Therefore, I must thank my Director, Dr. Kevin O'Connor, for offering the services of the MMB Breeding and Production Division, including the extensive photographic library. At the same time I should stress that the opinions expressed in this book are not necessarily those held by the MMB.

The majority of the photographs borrowed from the MMB were taken by their excellent Livestock Photographer Alan Bishop, whilst others were by colleagues Alan Carter, Martin Gee and Alex Park. My good friend David Abbott lent me transparencies from Finland, as well as his books on cattle. The remaining photographs were taken by Denis and Simon Bishop, by myself, or kindly supplied by one of the following:

Department of Primary Industries, Queensland (Australia)
Professor John Francis (Australia)
Mr Derek Froud (Netherlands)
Mr Erland von Hofsten (Sweden)
Mr Narayan Joshi (India)
Mr Les O'Hagan (Republic of South Africa)
The Luing Cattle Society Ltd (Scotland)
M. Pierre del Porto (France)
Snr Francisco José Simões (Portugal)
Dr. G Van Snick (Belgium)
South African Tourist Corporation
Mr Alan Sparger III (U.S.A.)
Snr Antonio Gonzales de Tánago (Spain)
Mr A. E. Timms (England)
Universal Livestock Services Ltd, Banbury (England)
Mr F. J. Willis (England)

Various Breed Societies have been most helpful and I would particularly like to thank Frank Pigney, Secretary to the English Guernsey Cattle Society for the loan of a valuable Herd Book. The American Brahman Breeders Association sent two photographs with their literature and the Texas Brangus Breeders Association forwarded a wealth of material.

To all my family, friends and colleagues, I thank you for your interest and help.

JOHN BENNETT FRIEND

ILLUSTRATOR'S ACKNOWLEDGEMENTS

Now that the book is complete, there are a number of people I wish to thank.

First of all, John Friend, who agreed to write the book, and did so, giving excellent information not usually found in other books.

Secondly, my son Simon, who has set a very high standard of colour illustration.

And finally my warmest thanks go to the men and women on the farms, who have allowed us to question them in minute detail, to photograph everything, and generally get under their feet. They have indeed given us more help than is usual nowadays.

I must also thank them for dedication to the job, which in turn provides the rest of us with a constant supply of dairy products, meat, and a cared for countryside.

Denis Bishop

PREFACE

All animals are fascinating, but for me cattle have a particular interest because of their very close association with humans. My own life has been closely linked with cattle breeding for a good number of years, so I was very pleased when the chance of writing this book came my way. Not everyone is fortunate enough to be able to share his or her interest with a wide audience.

Much research has been carried out on individual breeds of cattle and books are available that deal with cattle by countries or by continents. However, I have not yet come across a book that embraces cattle of the world in one volume, and I hope that this work will fill the gap.

The early chapters in this book cover the evolution of the cattle family, their domestication and the way in which they spread to all parts of the globe. This section precedes the breed descriptions and is intended to provide background information before proceeding to the different types of cattle that are the main subjects of the book. Most of the numerically larger breeds have been included as well as a selection of the rarer breeds. Where two or more breeds are very similar, but because of national boundaries have different names, they are included in the same description. This has enabled 111 breeds to be included under 90 main headings.

As it is very difficult to conjure up an impression of a breed by description alone, I felt it was essential to have an illustration to support the text. This became a limiting factor on the number of breeds included as it was not always possible to obtain a good photograph or satisfactory artist's impression.

At first glance the breeds may not appear to be in a logical order. I would, therefore, like to explain the reasons for the sequence. Following closely behind the Bison, Buffalo and Yak come the British breeds because they are some of the oldest established and best known worldwide. Next come the European breeds, as some of these are proving popular at present and are spreading to other parts of the world. Representative breeds from different parts of Africa and India are included before some of the newer American and Australian breeds. Finally, there is one of the very latest dairy breeds, the Jamaica Hope, followed by two new beef breeds.

Each breed description includes a few historical notes where there is general agreement on the origin of that particular type. Also included is an indication of what use the breed is to man, for very few breeds would survive today without human patronage. Although a certain amount of detail regarding working cattle, meat and milk animals is given in the breed descriptions, most of the information is common to a number of breeds. Therefore there are sections throughout the book, including Chapter 5, that discuss these uses more fully as well as other products from cattle. The many illustrations in the book are designed to enhance, and, in certain cases, supplement the text.

Throughout the time man has been connected with cattle he has tried to improve them in various ways. Chapter 6 indicates how rapidly progress can be made today with modern breeding techniques. In Chapter 7, I have tried to show what a breeder stands to gain by producing better cattle.

JOHN BENNETT FRIEND

1 THE CATTLE FAMILY

Cattle are classed as mammals because they have a backbone, hair and mammary glands. As this does not distinguish them from the hundreds of other mammals, naturalists have further divided them into Orders. In the order Artiodactyla—or cloven hoofed mammals—are pigs, camels, llamas, deer and cattle. These animals have two enlarged toes that form the split hoof, whilst the remaining toes are small and not used. Within this Order are a number of Families, and cattle belong to the family Bovidae. However, as this also includes sheep and goats, it is necessary to split the family into three sections and each section is known as a genus. Cattle are referred to as the genus *Bos*, which is the Latin word for ox. Within this genus are a number of species of cattle, including Bison, Buffalo, Yaks and their relatives, as well as the domesticated breeds.

Characteristics that the genus *Bos* have in common with the rest of the family Bovidae are horns which are never shed, a stomach with four compartments and mammary glands that are carried between the hind legs. Also, the dentition is different in cattle as they have no incisors or canine teeth in the upper jaw. The lower teeth at the front work against a dental pad on the roof of the mouth.

All modern cattle are thought to have evolved from a common ancestor that existed in Asia between two million and seven million years ago. This period of time is called the Pliocene and follows on from the Miocene, when the first archaic mammals were present on earth.

One of the first distinct species known to have developed in the Pliocene was *Bos primigenius*. The remains of various types of *Bos primigenius* have been found in India, Egypt and Europe and these animals were the forerunners of the domestic cattle that we know today. In Europe, *Bos primigenius* took the form of the now extinct wild 'auroch' described in the next chapter. It is believed that the humped cattle of India, *Bos indicus*, are descended from a type of *Bos primigenius* that has been called *Bos namadicus*.

Other members of the genus *Bos* are Bison, Yak, Gaur and Banteng; these can easily be interbred with domestic cattle, although the progeny are not always fertile. This would indicate that they evolved from the same prehistoric ancestor. Buffalo are not so closely related as they do not voluntarily interbreed with domestic cattle. Their ancestry is, therefore, uncertain, but fossil remains show that they were present as a distinct branch of the cattle family in the Pleistocene period, which began some two million years ago.

The Pleistocene is also known as the Great Ice Age, as much of northern Europe and North America were covered by great ice sheets. These receded and returned at intervals as the climate cooled again. Most of the present species of land mammals were in existence and were driven south by the glaciations. During this period, cattle were developing by the process of natural selection, and only those beasts that adapted to the harsh conditions were able to survive.

This evolutionary process has resulted in the formation of many distinct species of mammals; the following are present day members of the genus *Bos:*

Bos taurus the domesticated cattle of Europe

Bos indicus the domesticated cattle of India and Africa

Bos gaurus the Gaur of India and Burma

Bos banteng The Banteng of Burma, Java and Borneo

Bos grunniens the Yak of Tibet

Bos bonasus the wild European Bison

Bos bison the wild North American Bison

Bos bubalus the domesticated Indian Buffalo

Bubalus depressicornis the wild Anoa

Syncerus caffer the wild African Buffalo.

After the final retreat of the ice, the earth entered its most recent period, called the Holocene. Climates became warmer and as conditions were easier the process of human civilisation began. This started with the domination of other species and eventually led to the domestication of many animals, including cattle.

2 DOMESTICATION OF CATTLE

Cattle have been part of agricultural everyday life for thousands of years. A partnership has been built up where cattle supply the human race with a variety of products in return for protection and a guaranteed food supply. Therefore both partners benefit from the association although humans are the dominant species. The process of making animals dependent in this way is called domestication. To find out how, why and where cattle were domesticated, it is necessary to delve into antiquity.

Ancient History

Primitive man was a hunter who killed animals to obtain the necessities of life. Meat, hides, horn and bone from the wild cattle provided food, clothing, utensils and tools. From about 6,000 B.C. man began to change his way of life—from hunter to farmer—and started to domesticate the animals around him. First the dog, and then sheep and goats were brought into the mutual dependence of domestication.

When the cultivation of crops began, the wild cattle came and ate up what they could at every available opportunity. Most of these four-legged robbers were chased off and killed, but some were caught and brought into close contact with their human captors. A few calves might have been kept as pets, although most animals would have been treated as a ready supply of fresh meat.

As goats had been kept and milked for quite some time, it would only be a small step for primitive man to milk the cows he caught. The quantities of milk obtained would have been greater than from goats and therefore cattle would be considered a valuable addition to primitive farms.

Archaeological Evidence

Although the method by which cattle were domesticated is not known precisely, the timing of this event is known from archaeological evidence. One of the earliest civilisations to domesticate cattle was in the Middle East, where very old tomb pictures and statues depict cows being milked and butchers at work. A mosaic frieze from the temple of A-Anni-Pad-Da at Ur in what was Mesopotamia, now Iraq, shows cows being milked from behind, i.e. between the two back legs, just as goats are milked. This frieze dates from about 3,000 B.C. and a fragment of vase of the same period from Tell Agrab in Mesopotamia shows a humped bull standing in a stall. This and other evidence indicates that domesticated Indian humped cattle were moved into Mesopotamia as early as 4,500 B.C. Archaeologists have concluded that domestication in Asia and the Middle East must have taken place between 6,000 and 5,000 B.C. At this time there was a temperate climate in North Africa and Asia and vast areas were covered with rich vegetation, in contrast to the deserts that exist today. The flood plains of the Nile, Euphrates and Tigris rivers, with their good soils, were ideal for early agriculture and the keeping of cattle.

The area of domestic cattle spread as immigrants from the Middle East migrated south and west through Africa and then north and west into southern Europe. At the same time there are indications of a centre of domestication within Europe itself. In Denmark remains of domestic cattle have been excavated, and dated at 2,600 B.C.

Cattle in Early Art and Writings

It can have been no easy task for early man to tame the wild cattle, as the ancestors of todays quiet animals were really formidable beasts. Down through the ages they have had many names such as uri, urus, thur and auer but they are now commonly known as aurochs, or more scientifically *Bos primigenius*. Skeletal remains have been found in many

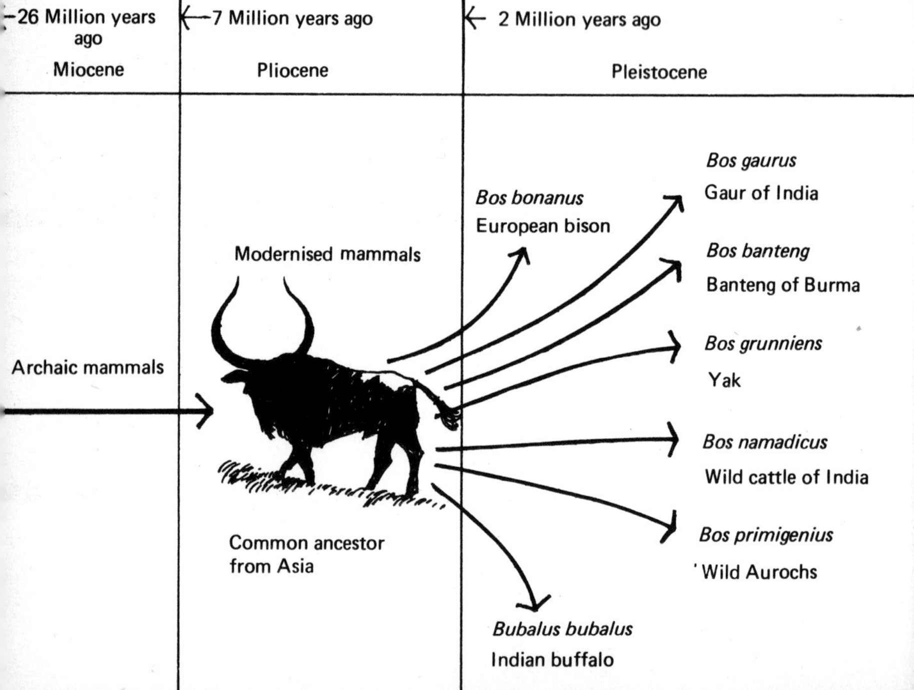

European countries across to Russia, Western Asia and northern Africa.

The earliest pictorial records of aurochs come from the Lascaux cave paintings in south-west France, dated 15,000 B.C. These show black or reddish-black bulls with a light coloured line along the back. A few white cattle are shown and some of these have black spots on the head and front part of the body. Further visual evidence is carved on the mortuary temple of Ramases III at Medinet Habu. One scene shows this Egyptian king hunting in the reed marshes with bow and arrow. There are three dead, or dying, wild bulls in the picture.

Julius Caesar compared the aurochs to elephants when writing in 65 B.C. and although he exaggerated their size he accurately recorded their strength, speed and ferocity. In A.D. 77 Pliny the elder, a naturalist of that time, drew attention to the aurochs 'remarkable strength and swiftness'.

The Aurochs

The Emperor Charlemagne hunted the wild aurochs in the forests near Aix-La-Chapelle in the 9th century, but by about 1400 they were extinct in western Europe. They survived in eastern Europe for some while longer, until the last one died in Poland in 1627.

From the various descriptions, artists impressions and archaeological remains, it is possible to get some idea of what an auroch looked like. Natural selection would have played a part in forming different types in different geographical areas. After domestication, preferred types would have been selected and in

Timetable Showing Domestication

	←10,000 years ago	←5,000 years ago
Pleistocene	Holocene	Beginnings of domestication
Bos namadicus		*Bos indicus* Domesticated zebu or humped cattle of India
Bos primigenius		*Bos primigenius* Domesticated humpless cattle of Europe and Asia

particular the smaller, more manageable animals would have been chosen.

The aurochs were big when compared with modern cattle. They were long-legged with adult males standing 175–200 cm (69–79 in) at the withers; cows were slightly shorter. The horns were big and lyre-shaped and on some skulls that have been excavated they measure almost 1 m (3 ft) from tip to tip. Very strong, muscular necks were needed to support the heavy skull and horns. Coat colour of the males was probably black and certainly darker than the brownish-red of the females. Both sexes usually had a white stripe along the back.

Reason for Domestication

What was there to gain by domesticating these large, wild aurochs? Once tamed they had to be fed and cared for. Early man already had goats to provide milk and wild cattle were plentiful and could be hunted for meat and hides. The answer probably lies in their strength. Any farmer today will tell of the power required to cultivate land and haul heavy loads. As the human race became farmers a source of energy was needed to carry out all the farming operations. There are still areas of subsistence farming in the modern world where cattle are used for draught purposes. Cattle were the first beasts of burden and later on the Elephant, Camel and Horse were tamed for the same purpose.

Not all types of cattle were domesticated. The Bison of Europe and America were too wild, as were the African Buffalo. Indian Buffalo were tamed and have proved invaluable as draught animals. The Yak is another beast of

Evolution of Separate Cattle Types

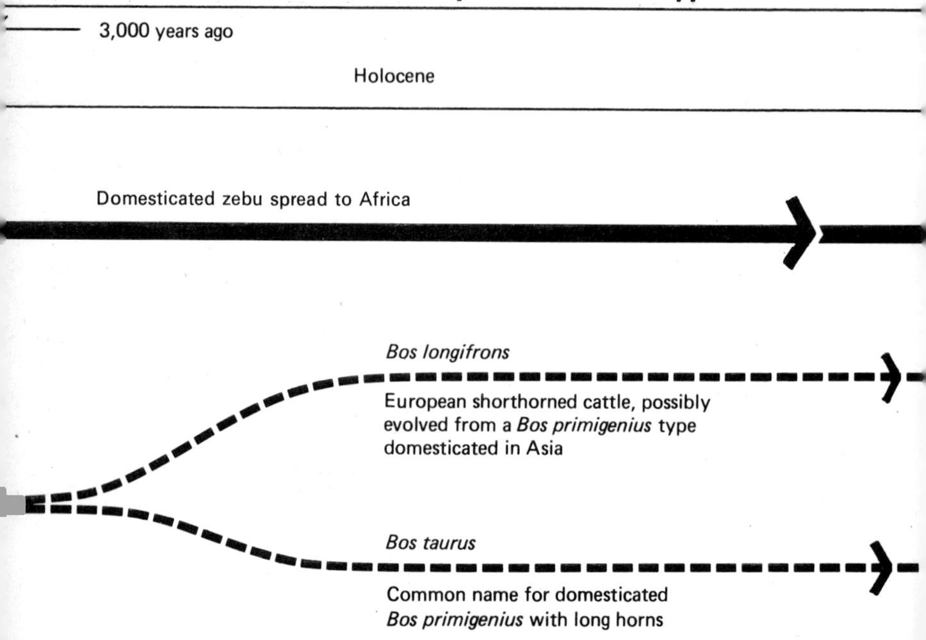

3,000 years ago

Holocene

Domesticated zebu spread to Africa

Bos longifrons

European shorthorned cattle, possibly evolved from a *Bos primigenius* type domesticated in Asia

Bos taurus

Common name for domesticated *Bos primigenius* with long horns

burden that has repaid man for the trouble of domesticating such a temperamental animal.

3 CATTLE MIGRATIONS

Cattle have spread all over the world, but the achievement of this situation has taken millions of years and considerable human assistance. The story begins some six million years ago in the Pliocene, when the ancestors of our modern mammals evolved from the archaic mammals present in the earlier Miocene (see map on pages 20–21).

The Beginning

The common ancestor of the cattle family began life in Asia and gradually spread to Africa, Europe and North America. During this long period, different varieties of cattle were formed and adapted by natural selection to their new habitats. Banteng, Bison, Buffalo, Gaur, Yak and the wild ancestors of today's modern breeds of cattle all went their separate ways.

Not all the main land masses were colonised by wild cattle. Bison managed to reach North America in the Pleistocene but did not get to South America. Australia was never inhabited by the early wild cattle. The Buffalo remained in Asia and Africa, whilst the Yak, Gaur and Banteng did not move far from their Asian homeland.

Man-made Migrations

The next stage in the spread of cattle did not occur until mankind began to

Timetable of Breed Development

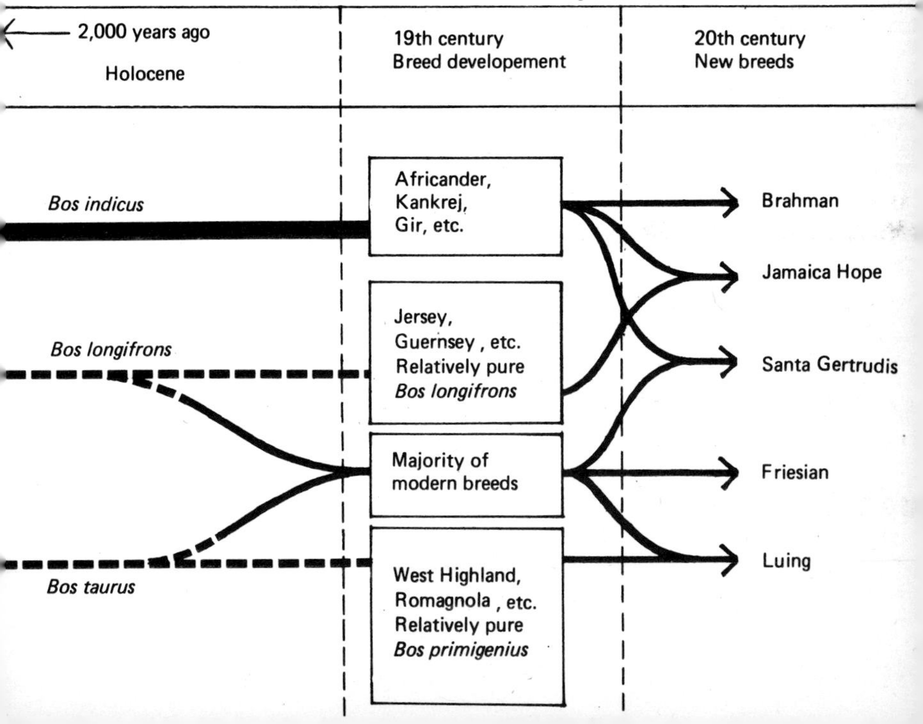

← 2,000 years ago
Holocene
19th century
Breed developement
20th century
New breeds
Bos indicus
Africander,
Kankrej,
Gir, etc.
Brahman
Jamaica Hope
Jersey,
Guernsey, etc.
Relatively pure
Bos longifrons
Bos longifrons
Santa Gertrudis
Majority of
modern breeds
Friesian
Luing
Bos taurus
West Highland,
Romagnola, etc.
Relatively pure
Bos primigenius

domesticate the animals. Three early centres of domestication of cattle have been found by archaeologists in India, the near East and Egypt. These occur in the fertile river valleys of the Indus, Tigris and Euphrates, and the Nile. It is thought that domestication in these areas occurred between 6,000 and 4,000 B.C.

In India, the Buffalo was domesticated alongside the *Bos indicus* or humped zebu cattle. Both types were used for draught purposes and played an important role in the early Indian civilisations. In the valley of the Indus river, in what is now Pakistan, some very sophisticated cities existed around 2,000 B.C. These had trading connections with Mesopotamia and Egypt and cattle had been exchanged for 1,000 years or more. *Bos indicus* cattle were bred alongside the humpless *Bos primigenius* cattle in these areas and greatly improved strains were produced.

Some of the first cattle to be domesticated in Africa were the giant horned wild oxen of the Nile Valley. Many of these animals were black and white or fawn and white—although blacks, whites, fawns and roans also appear on Egyptian wall paintings. The descendents of these now extinct Hamitic Longhorns are the Kuri of West Africa, the Ankole and Nguni cattle of East Africa, and the Nilotic cattle of the Sudan.

Migration to Europe

Smaller, shorthorned cattle are also illustrated in Egyptian sculptures and tomb paintings. These were the ancestors of the Egyptian, Libyan and Brown Atlas cattle found along the North African coast today. A migratory route can be traced along to Morocco, where the route divided. Some cattle continued down the coastal belt of West Africa and met up with the longhorn types. Here, crossbreeding produced the N'Dama stock. Other animals were taken by tribesmen across to Europe where their paths can be traced by the locations of their descendents. The Galician Blond in Spain and the Blonde D'Aquitaine in France point to the fact that the migratory route crossed the Pyrenees before it diverged. One path headed northwards across France to the Channel Islands where the Jersey cattle count among their modern relations. The second path crossed southern France, leaving behind the ancestors of the Aubrac and Tarentaise breeds, before reaching Switzerland, where the Brown Swiss can be found today. All these breeds have a high forehead and have been given the collective name of *Bos longifrons*.

Another important route to Europe for tribesmen and their cattle was through south-west Asia, across the Bosporus and then heading west. One group of Neolithic lake dwellers reached Switzerland with domesticated cattle. They settled there and proceeded to domesticate the local wild *Bos primigenius*. Excavations of sites around Lake Zurich have shown remains of three different types of cattle. The largest of these are the wild *Bos primigenius*, whilst the smallest are considered to be the cattle brought to the area by the lake dwellers. Intermediate types are thought to be crosses between the two.

During the many hundreds of years that it took for people and cattle to spread out from the early centres of domestication, European man was also taming the wild oxen. There is archaeological evidence to show that Denmark and the Schleswig-Holstein area of Germany was one such centre of early domestication in Europe. Probably from around 3,000 to 1,500 B.C. the wild *Bos primigenius* type in this area was tamed.

The Great Trading Nations

The spread of cattle considerably increased at the time of the great trading civilisations around the Mediterranean Sea. The Etruscans, Greeks and Phoenicians all played a part and traded with the barbarian Celtic races in northern Europe, as well as the peoples around the Mediterranean. Sea routes were used as men and their goods could travel quicker this way than overland. Then came the rise of the Roman Empire, which after conquering nearly all the countries with a Mediterranean coastline, turned northwards against the Celtic world. By A.D. 43 Britain was occupied by the Roman legions, and so was that part of Europe to the south and west of the Rhine and Danube rivers. The Romans took their cattle with them and the Charolais breed in France is thought to have descended from animals left behind by the Roman legions. In Britain, the Wild White Park Cattle are very similar in markings to the white Italian breeds and probably relate back to the cattle that were allowed to run wild after the Romans departed.

Europe after the Romans

When the Roman Empire collapsed, various races of people competed for the territory that was once occupied by the legions. The British Isles, for example, was repeatedly invaded and each time the invaders introduced their own type of cattle. The Anglo-Saxons brought with them red cattle, whose descendents exist today as the Devon and Sussex breeds. When the Vikings invaded Britain, they had large cattle that were the early ancestors of the Lincoln Red breed, and hornless cattle that introduced the polled factor into Britain. Viking raiders also landed cattle in northern France and the Normandy breed is thought to have descended from these animals.

In southern Europe another migration was taking place which spread the longhorned Steppe cattle from the east to Italy, the south of France, Spain and Portugal. Related modern breeds are the Romagnola in Italy and the Gasconne in France. More recent movements of cattle in Europe are dealt with under the respective breeds.

Cattle of the New World

No domesticated cattle existed on the American continent until the Spanish conquerors brought their animals with them early in the 16th century. North America had Bison, which the Indians had found impossible to tame, and South America had no cattle at all. The Texas Longhorn, made famous by cowboy films, is the living descendent of those first Spanish cattle. Many European cattle came to America in the wake of the Spaniards. Dutch settlers landed animals in 1621 and English and French colonists followed suit.

Australia was known to Portuguese explorers in the 16th century and the Spanish, Dutch, French and English had all made landings by the end of the 17th century. However it was not until January 18th, 1788, that the first

English colonists landed at Botany Bay. As there was no wild or domestic cattle in Australia until the first real colonies were started, the settlers imported animals from England. European breeds of cattle are now widespread in Australia, South America and North America.

Spread of the Zebu

Until recent times the humped 'zebu' cattle of India had not spread further than Africa and parts of Asia. Then, in the mid-19th century, they were imported into both North and South America. Brazil took quite a number and so did the southern states of the U.S.A., where the cattle breeders developed the American Brahman. These Indian animals proved popular with the ranchers as they could withstand the hot, humid weather and repel insect pests.

Early in this century the Australian property owners in the hot north of their country also saw the advantages of 'zebu' over the European breeds. The 'zebu' survived much better in the hot, dry conditions and the Droughtmaster breed in Australia is a 'zebu' crossed with Shorthorn.

Modern Migrations

In the 20th century, with improved forms of transport, it has been possible to move cattle easily all around the world. Holstein-Friesian cattle from Canada and America have been transported to Europe, Australasia and South America. The beef breeds of Europe, such as Charolais and Simmental, are now being bred in North America, and crossed with Indian 'zebu'.

Two technical breakthroughs have

Cave drawing of wild bull, Lascaux, South-West France.

Tomb painting of Thebes, Egypt, showing a bull.

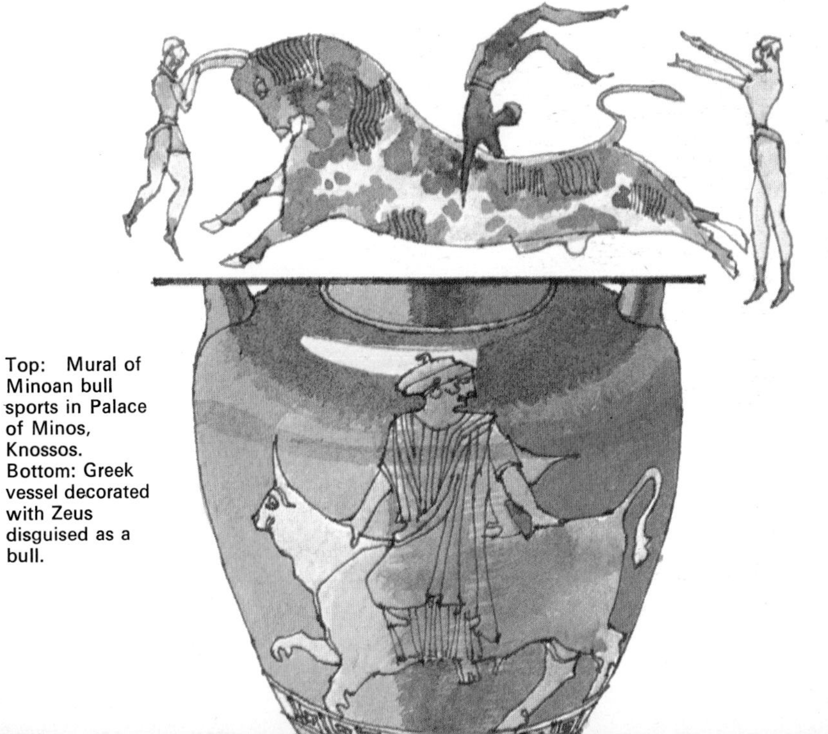

Top: Mural of Minoan bull sports in Palace of Minos, Knossos.
Bottom: Greek vessel decorated with Zeus disguised as a bull.

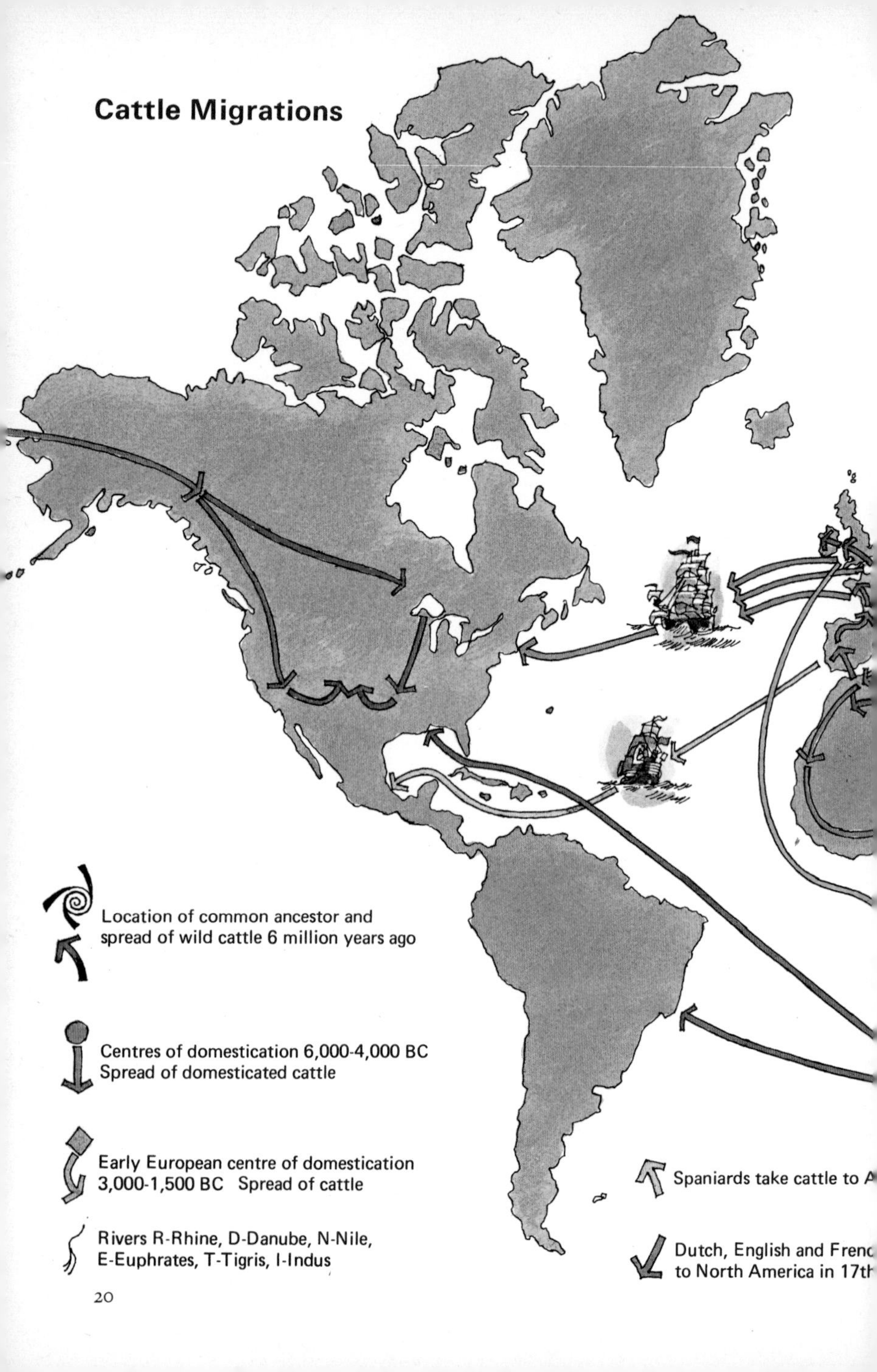
Cattle Migrations
Location of common ancestor and spread of wild cattle 6 million years ago
Centres of domestication 6,000-4,000 BC Spread of domesticated cattle
Early European centre of domestication 3,000-1,500 BC Spread of cattle
Rivers R-Rhine, D-Danube, N-Nile, E-Euphrates, T-Tigris, I-Indus
Spaniards take cattle to A
Dutch, English and Frenc
to North America in 17t

British cattle to Australia at end of 18th century
ica in 16th century
Indian Zebu cattle to America and
Australia, mid 19th century
lonists take cattle
tury
Constant movement of cattle throughout middle east

Greek drinking cup showing a dead Minotaur that was half man and half bull.

Terracota bull of the late Minoan period from Rhodes.

revolutionised the cattle breeding world and reduced the need to ship or fly live animals. The first is artificial insemination and the freezing and storage of semen for long periods. Thousands of doses of semen can be transferred from country to country in liquid nitrogen storage flasks for the cost of moving a single animal. As long as the country importing the semen has the technicians to inseminate the cows, then it can use genetically superior bulls from anywhere in the world.

The second breakthrough was the perfection of the ova transplant technique, where fertilised eggs from one cow can be implanted in a recipient cow.

These innovations are discussed in Chapter 6, 'Modern Breeding Techniques.'

Oxen harness from Gloucestershire, England.

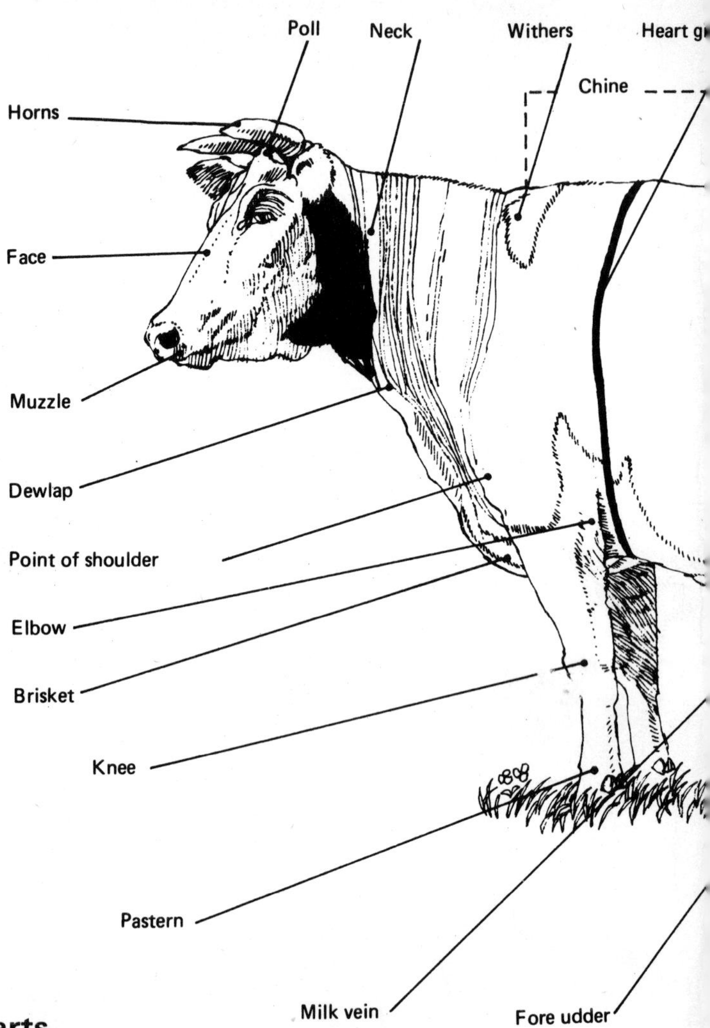

Cow Parts

Breeders of cattle pay particular attention to the appearance of their animals, which they call 'type' or 'conformation'. Each part of a cow or bull has a name which cattlemen use when discussing their animals. These cow parts, or points of conformation, are shown in the above diagram which may be helpful when reading the breed descriptions.

Although breeds differ, there are basic conformation characteristics that

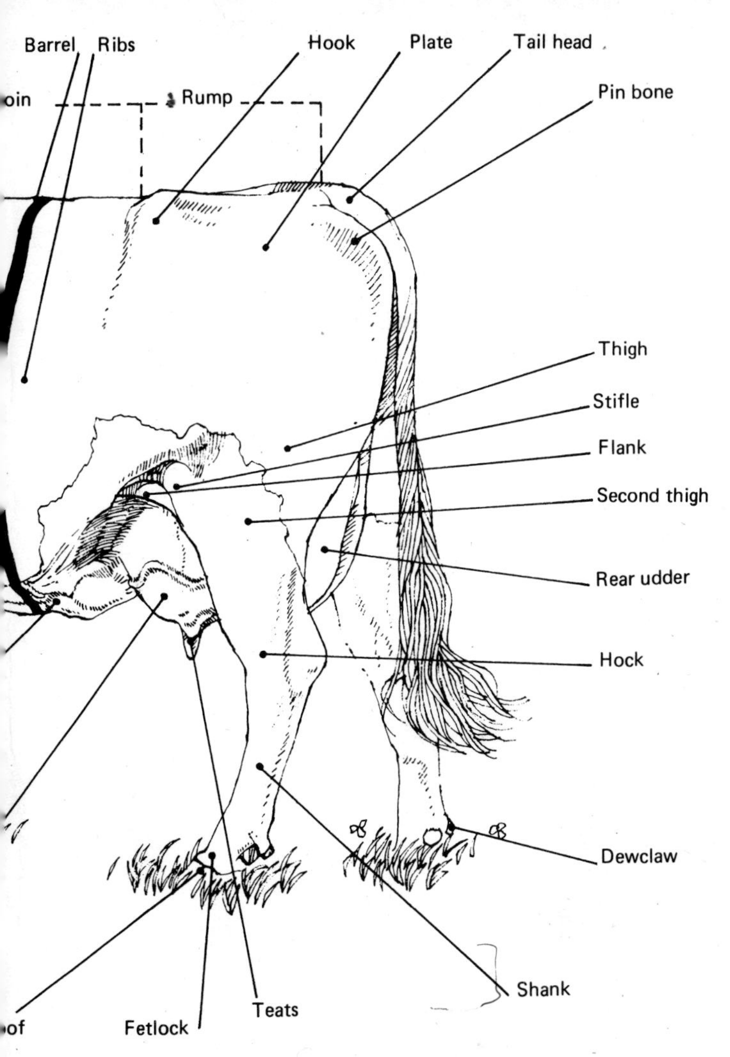

all breeders concentrate on improving. With dairy cows, the udder is very important and must be attached to the body with strong muscles. The teats must be evenly spaced, not too big or too small, and should point straight down towards the ground. Beef animals should be long and well muscled, particularly over the rump and down the thighs.

4 CATTLE BREEDS

A description of all the following breeds can be found in this extended chapter.

Apart from those listed in smaller type, all are illustrated in colour – except for the British Friesian, which is a black and white breed.

Breed	*Country*
European Bison	Europe
American Bison	North America
African Buffalo	Africa
Cape Buffalo	South Africa
Congo Buffalo	West Africa
Dwarf Buffalo	Indonesia
Indian Buffalo	India
Gaur	India, Burma, Malaya
Gayal	India
Banteng	Burma, Malaya, Java
Yak	Tibet
Wild White Park Cattle	England
British White	England
Longhorn	England
Old Gloucestershire	England
Welsh Black	Wales
West Highland	Scotland
Galloway	Scotland
Belted Galloway	Scotland
Aberdeen Angus	Scotland
Hereford	England
Devon	England
South Devon	England
Sussex	England
Lincoln Red	England
Ayrshire	Scotland
Shorthorn	England
Beef Shorthorn	Scotland
Guernsey	Guernsey
Jersey	Jersey
Kerry	Ireland
Dexter	Ireland

Breed	*Country*
Black and White Dutch Friesian	Netherlands
Holstein-Friesian	U.S.A., Canada
British Friesian	Great Britain
Red and White Friesian	U.K., Canada, U.S.A.
Normandy	France
Tarentaise	France
Aubrac	France
Salers	France
Charolais	France
Limousin	France
Blonde d'Aquitaine	France
Maine Anjou	France
Gasconne	France
Brown Swiss	Switzerland
Simmental	Switzerland
Fleckvieh	Germany
Pie Rouge de l'Est	France
German Red Pied	Germany
Angeln	Germany
German Yellow	Germany
Pinzgauer	Austria
Chianina	Italy
Marchigiana	Italy
Romagnola	Italy
Piemontese	Italy
Meuse-Rhine-Ijssel	Netherlands
Groningen White-headed	Netherlands
Central and Upper Belgian	Belgium
Danish Red	Denmark
Swedish Red and White	Sweden

Breed	Country
Telemark	Norway
Blacksided Trondheim and Nordland	Norway
Finncattle	Finland
Finnish Ayrshire	Finland
Kholmogor	Russia
Red Steppe	Russia
Ala Tau	Russia
Kostroma	Russia
Mongolian	Mongolia
Miranda	Portugal
Barrosa	Portugal
Galician Blond	Spain
Galega	Portugal
Andalusian	Spain
Fighting Bull	Spain
Africander	South Africa
Drakensberger	South Africa
Nguni	South Africa
Landim	Mozambique
Bapedi	South Africa
Boran	Kenya
Ankole	Uganda
Bahima	Uganda
Watusi	Tanzania

Breed	Country
Bashi	Zaire
Kigezi	Uganda
Kuri	Chad
White Fulani	Nigeria
N'Dama	Guinea
Egyptian	Egypt
Saidi	Egypt
Maryuti	Egypt
Damietta	Egypt
Baladi	Egypt
Kankrej	India
Gir	India
Khillari	India
Tharparkar	Pakistan
Red Sindhi	Pakistan
Texas Longhorn	U.S.A.
Brahman	U.S.A.
Santa Gertrudis	U.S.A.
Brangus	U.S.A.
Droughtmaster	Australia
Murray Grey	Australia
Tasmanian Grey	Australia
Jamaica Hope	Jamaica
Luing	Scotland
Beefalo	U.S.A.

European Bison

The European Bison has fortunately been saved from extinction but can now only be found in zoos and reserves. It once inhabited most of the forests in Europe and North Asia. In the 15th century it was so numerous that stampeding herds were a hazard to the peasants in these regions. In the 17th century the European Bison was being extensively hunted in Poland. From this position of strength the breed was brought to the point of extinction.

During World War I, these beasts were slaughtered for food in the areas they still frequented and were driven from their natural habitats. The really wild race in the Caucasus practically died out and all that remained in Europe were a few small herds that had been preserved in the Lithuanian forests.

This, the largest of all European quadrupeds, was also known as the *Wisent* (German) or *Zubr* (Russian). It is estimated that naturally grazing animals require 40 hectares (about 100 acres) of forest each to support them.

A thick coat of short, curly brown hair covers most of the body, while the head, neck, shoulders and fore limbs are clothed with a mane of longer, darker hair, Bison have massive heads with broad convex foreheads. Their shoulder hump is made up of the large muscles required to support the head and also longer bits of bone that stick out from the spine at this point. Another characteristic is the number of ribs. In common with the Yak, Bison possess fourteen pairs of ribs while ordinary cattle have only thirteen. The horns are comparatively small and slender with an upwards curve.

American Bison

There were once great numbers of Bison roaming over vast areas of North America where they were the only native bovine. One estimate put the figure as high as 60,000,000. Although the great western prairies were their natural home, they did not just confine themselves to this open country. Bison were to be found from the Atlantic coast across to the Pacific and from northern Mexico right up to the shores of the Great Slave Lake in Canada.

The American Bison, or 'Buffalo' as it was wrongly called, provided the Red Indian tribes with the necessities of life, even though it could not be tamed. The rich, nourishing meat was the Plains Indians' staple food. Hides provided clothing, cooking pots, shields, ropes and shelter. Between ten and forty skins were used to make a teepee, depending on how large the home was to be. Horns, hooves, bones and sinews were made into tools, weapons, glue and thread.

For hundreds of years the Indians lived on the Bison without seeming to reduce their numbers, in spite of the large slaughterings that took place at the time of the spring migration. Every winter the Bison moved south to fresh pastures and in the spring set off northwards again along regular routes.

When the white man came the balance was upset as this 'protein on the hoof' was hunted continuously to provide food for the settlers pushing ever westwards. Between 1730 and 1830, the Bison were driven away from the Eastern U.S.A. and also from the country to the west of the Rocky Mountains.

From 1830 onwards, the spread of civilisation sealed the fate of the Bison. There was systematic slaughter for meat and hides, and in some instances just for the tongues to supply canning factories. By 1883, the great herds had gone forever. Some 300 animals settled in the Yellowstone National Park under the protection of the United States Government and larger numbers survived in Canada where they can be seen in nature reserves.

The American Bison is a darker brown, has heavier forequarters and a more pronounced hump and mane than the European variety. The horns are short, blunt and curved, differing from the European Bison in being set further back on the head. Cows usually calve from April to the end of June. The calf's first coat is a chestnut colour. Adults shed their winter coat in June each year and can have an almost naked back at this stage. To prevent sunburn and discourage flies the animals roll in the mud which forms a thin protective layer.

Bison will breed with ordinary cattle, although the hybrid produced is not as fertile as either of the parents. Recently a new breed has been launched in North America. The Californian rancher who has developed this hybrid calls it a Beefalo, and it includes $\frac{3}{8}$ Bison, $\frac{3}{8}$ Charolais and $\frac{1}{4}$ Hereford (see page 157).

African Buffalo

There are a number of varieties of the African Buffalo spread all over the African continent. None have been domesticated and they have been there for centuries. Archaeological evidence is to be found at Ksar el Ahmar in North Africa where there are Neolithic rock engravings of these beasts.

The largest variety is the Black Cape Buffalo and it is considered the most dangerous animal in Africa. Hunters used to fear the charge of this beast more than any lion. Mature males can be 274 cm (108 in) long and carry massive horns that are broad at the base and nearly meet in the middle of the forehead. They have large, flapping ears, with a thick fringe of hair. The body hair is very thin and older animals can be almost bare.

The Congo Buffalo of West Africa and the Congo river basin is much smaller with upturned short horns and bright reddish hair.

African Buffalo tend to feed at night and rest by day. However, it has been noted that where lions roam at night the buffalo have changed to feeding by day.

Buffalo can carry in their blood parasites that cause sleeping sickness. Although they are not affected by the disease, they act as carriers from which the tsetse fly can transmit it to humans.

Dwarf Buffalo

The Anoa, or Dwarf Buffalo is found wild on the island of Celebes. It does not grow to much more than 91 cm (36 in) in height and resembles a short-legged antelope rather than a Buffalo. Although shy and retiring, it can be dangerous when captured and has never been domesticated. The natural habitat of the Anoa is remote wooded areas and it is very rarely seen.

These Dwarf Buffalo are usually black in colour, with some white markings on the face, neck and fetlocks, although some females are brown. They have a neat head with sharp, backward pointing horns that grow to about 30 cm (12 in) and have a triangular base. The low carriage of the head and the long heavy body are buffalo characteristics and in fact this breed resembles the young of the Indian Buffalo.

Indian Buffalo

The Indian Buffalo is a native of India and Sri Lanka. It is also called Arnee and Water Buffalo, as in the wild it lives in grass jungles near to water. Many hours of each day are spent wallowing in the water or just lying submerged except for nostrils and horns. Although a most dangerous beast, attacking even when unprovoked, it has been successfully domesticated.

Buffalo are most useful as draught animals, particularly in rice growing countries as they can work in water and deep mud. They can each plough up to 0.1 hectare (0.25 acre) in a six hour day in a paddy-field. A single rein is used to guide them and this is attached to a nose plug which is permanently inserted in young animals. Other tasks include the working of mills and the pumping of water from wells.

Milk from the Buffalo is much richer than ordinary cow's milk and can be made into solid butter. The greenish-white colour of this butter is not very attractive but it is ideal for hot climates as it keeps very well. The dark Buffalo meat is eaten and the skin makes exceptionally heavy and strong leather.

Domestication of the Indian Buffalo meant that it spread both east and west from its original home as other countries realised how useful it was for work. In China it has provided power on the farms for some 3,000 years and it is estimated that there are some 30,000,000 at present in the Chinese Peoples' Republic.

Although the Buffalo spread eastwards very early, it made much slower progress in a western direction. It did not reach Egypt until the Middle Ages and then did not spread any further into North Africa. By A.D. 1200 Bulgaria and Macedonia had thriving Buffalo populations and from here they spread into Hungary. They were also established in Russia, Italy and Sicily but attempts to introduce them into France, Germany and Spain were failures. With cattle supplying all basic needs there was no real incentive to experiment with the much less docile Buffalo.

The Indian Buffalo has a thin coat of black or reddish-brown hair, which it loses with age. Lighter colours and even albinos are found amongst the tame population. Horns are large, flat above, strongly ribbed and curve backwards. Many different types of Buffalo have been bred since domestication. They can be distinguished from each other by body size, length and shape of horns, and the amount of hair.

Water is essential for the Indian Buffalo and working beasts are allowed rest periods during the day for wallowing, which prevents their skin drying out and cracking. If there is no mud, then each Buffalo digs its own wallow in the ground with horns and forefeet until water seeps in and fills the hole. The food they eat is much coarser than

European Bison (Europe)

American Bison (North America)

Cape Buffalo (South Africa)

Dwarf Buffalo (Indonesia)

Indian Buffalo (India)

Yak (Tibet)

Gaur (India, Burma, Malaya)

Banteng (Burma, Malaya, Java)

that of cattle as they will tackle reeds and other swamp vegetation.

Cows do not breed until they are at least three years old and their gestation period is a little over ten months. Animals of twenty to twenty-five years are common, and some are even known to have working lives of thirty years.

Buffalo are extremely insensitive to pain and even quite serious injuries do not appear to cause much loss of condition. Biting flies which would stampede cattle are ignored and whips are practically useless as a means of discipline.

Gaur and Gayal

The Gaur is the largest of the wild oxen still in existence. Mature bulls can reach a height of 183 cm (72 in) at the shoulder and achieve a length of 290 cm (114 in) This animal, also referred to as the Indian Bison, inhabits hilly forests of India, Burma and Malaya. It has nocturnal habits, feeding during the night on grass and young shoots of trees and bamboos, and retreating into the depths of the jungle during the day. Although naturally shy and retiring it can be a formidable opponent when cornered, and hunting these beasts used to be a favourite sport of Anglo-Indians.

The general colour of the Gaur is black or dark brown, with the lower part of the legs being white. The short coat consists of very fine, glossy hair. Characteristic features of this breed are massive curved horns somewhat flattened at the base where they come out from the forehead with its convex ridge. Behind the head is a shoulder hump due to the structure of the backbone. Each vertebrae from the third to the eleventh has a long bony process. Cows are much smaller than the bulls and usually calve in August and September.

A close relative of the Gaur is the Gayal which has become domesticated and is smaller and has shorter legs. The horns of the Gayal are thicker and flatter than those of the Gaur, and it has a great bony forehead.

Although tame, the Gayal does not have a great deal of use and is not employed very much for agricultural work. However, the meat is eaten, the hides are used for leather and the beasts are sacrificed in ceremonies performed by the Naga tribesmen of Assam. Gayals vary in colour and are often white or yellow. Animals prized for sacrifice are normally black like the wild Gaur and are selectively bred for the size and span of their horns.

The Gayal is interbred with local 'zebu' cattle and the offspring are favoured in certain districts. Male animals resulting from these crossings are infertile.

Banteng

The Banteng is found in Burma, Thailand, Malaya, Java and Borneo and has been domesticated in this area. As well as being a working animal, it supplies meat for the population, and is often mated with the local 'zebu' type cattle.

Banteng in the wild live in small herds, feeding in the early morning and then retiring into the jungle for shelter. In Burma the local name for the Banteng is *Tsine* and this variety lives in the grassy plains and leaves the hill-forests to the larger Gaur. The *Tsine* are all a bright reddish-brown, but mature bulls of other types of Banteng turn black. All have a large white patch on their rumps and white stockings from above the knees down to the hooves. The tail comes down well below the hocks and the horns vary in size and shape with the different types. A similarly-formed shoulder hump is present in the Banteng as in the Gaur.

Yak

Tibet is the home of the Yak and it has been a domestic animal in that country for centuries. Because it can survive on the wiry grass and withstand the freezing temperatures in this part of the world, it is an invaluable 'beast of burden'. Loads of 150 kg (330 lb) are carried with ease over dangerous mountain paths and the Yak is essential for journeys at high altitude in Central Asia.

Although their main function is for work, the Yak can supply many other needs in this desolate area of high plateaux and mountains. The rich Yak milk is made into butter and an alcoholic drink, as well as being the main ingredient of *tsamba*, a concoction with tea and barley meal. The hair of the Yak is used for clothing and as there are no trees in the Tibetan highlands the dung is burnt as fuel. The meat is eaten and even the tails were once sought after by Indian Princes for use as fly whisks.

Wild Yaks are always black, but domesticated animals vary from black, through brown, to white. They are long and low with long, fine, glossy hair that falls off along the side into a sweeping fringe. Horns are large and black and curve upwards and forwards in the male. The head is carried low and the neck leads into the powerful humped shoulders.

The Yak has a number of peculiarities including its strange grunting cry. It has a very keen sense of smell which makes it difficult to get near when in the wild in spite of its poor sight.

Yaks will not eat corn and when grazing only feed in the mornings and evenings, preferring to lie down and rest during the day. As they will eat only naturally growing food and suitable pastures can be thirty or more miles apart this is a great inconvenience to their owners.

It is possible to cross breed Yaks with ordinary domestic cattle including 'zebu'. The product of a male Yak and a female 'zebu' is called a 'Zo', whilst a male 'zebu' on a Yak cow produces a 'Zomo'. These hybrids are not so strong as the purebred Yak. Another disadvantage is that the males are infertile. However, they are popular as they are quieter animals and can be used for ploughing, a task which the Yak performs only very reluctantly.

In Britain recently, a Lincolnshire farmer has bred what he calls a 'Yakow' by crossing a Highland cow with a Yak. The intention is to create a type of hardy animal that will produce lean meat on a scrubland diet.

The Yak, the traditional beast of burden in Tibet

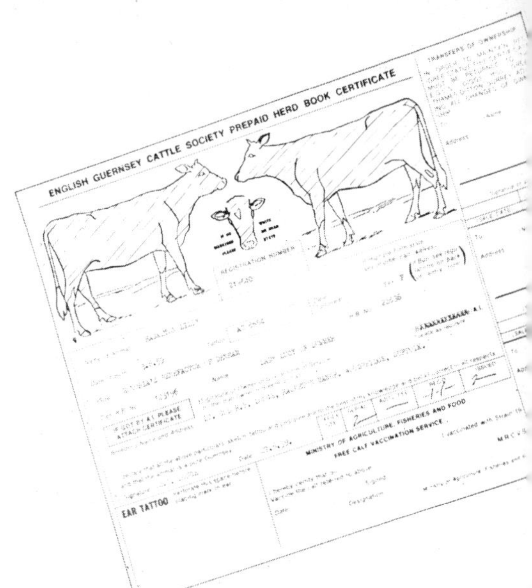
ENGLISH GUERNSEY CATTLE SOCIETY PREPAID HERD BOOK CERTIFICATE

REGISTRATION NUMBER

MINISTRY OF AGRICULTURE, FISHERIES AND FOOD

FREE CALF VACCINATION SERVICE

EAR TATTOO

TRANSFERS OF OWNERSHIP

Breed Societies and Herd Books

Cattle breeders are farmers, who by controlling the matings of their animals, attempt to improve the livestock that they own. Down through the years, breeders interested in producing the same type of cattle have grouped themselves into Breed Societies. The aim of each Breed Society is to promote its own type of animal both at home and abroad.

Every breed society has its own agreed rules that members must adhere to and which are designed to safeguard the future of the breed.

Breed Society members pay a subscription which meets part of the cost of running an office and employing a Breed Society Secretary and other staff. The permanent office is essential because of all the paper work involved in running a thriving Breed Society. A membership register has to be maintained and a Herd Book kept for all animals registered with the Society. Most Societies publish their Herd Book annually so that members can trace the ancestry of pedigree animals. Promotional material such as Breed Society Journals and advertising literature are also circulated.

Herd Books contain the following information on every animal entered, or registered, with the Breed Society: Name; Date of birth; Herd Book Number; Names and Herd Book Numbers of Sire and Dam. All these details, except the Herd Book Number of the animal, are supplied by the breeder to the society after the birth of a calf. The information is sent in on an official registration certificate, and accompanied by the appropriate fee. Some societies require a sketch of the calf showing a pattern of the markings. This is particularly useful in the case of Friesian and Guernsey breeds where no

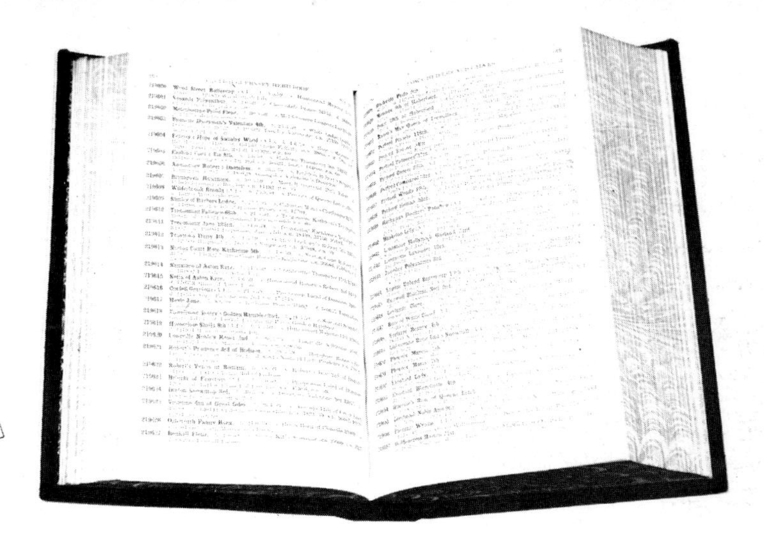

two animals have the same markings. Other breeds insist on tattoo marks being placed in the ears of the cattle. If the registration is accepted by the Society, then a Herd Book Number is issued.

A number of Herd Books are 'closed' which means that only animals whose parents are entered in the Herd Book are eligible for registration. The remainder are 'open' and admit animals, provided they meet certain requirements of colour and type, even if neither parent is registered. Where 'open' Herd Books are kept, it is possible to import a similar type of animal from another country and register it in the Herd Book of its new owner's country.

Animals registered with Breed Societies are called *pedigree*: those not registered are 'commercial' or non-pedigree stock. It is possible with some 'open' Herd Books to 'grade-up' females from non-pedigree to pedigree under the Society's rules. The process normally takes five generations using a pedigree bull on a grading cow each time. The grading animals are entered in a separate register, or supplement to the Herd Book during the transition period. Some societies arrange for the grading cattle to be inspected to see if they are true to the preferred type before entering them in the grading up register.

In addition to the regular Herd Book, many breeds have an Advanced Register or Production Register. This is a separate book in which the performance of outstanding animals is recorded. This type of publication acts as an incentive to the breeder; encouraging him to strive for better animals.

Wild White Park Cattle

The history of this breed goes back into the unrecorded past. Possibly these cattle are descended from the white bulls that featured in Druid sacrifices in ancient Britain. More likely they were the remnants of herds brought over by the Romans and left to run wild when the Roman occupation ended. In colour markings they do resemble the Italian Chianina cattle.

During the 13th century, these wild cattle were rounded up and driven into parks where a few of the original herds still survive today. The oldest and purest of the herds is at Chillingham Park in Northumberland, where in 1220 a wall was built to enclose the 120 hectares (300 acres). From that day to this no outside blood has been introduced.

Chillingham cattle are still wild and the bulls fight each other to see which one will rule the herd and sire the next crop of calves. The cows seek a secluded spot in which to give birth and then stay with their new born for between four and fourteen days. If at the end of this time the calves are not strong enough to run with the herd they are then abandoned. Sick animals are exiled or killed but fortunately common ailments such as digestive troubles, mastitis, worms and warbles are unknown. The herd numbers about 50 at present with 23 males and 28 females.

In the north of England and Scotland, Wild White bulls were hunted for sport until some 300 years ago. There was some organised hunting at Chillingham Park as late as the early 19th century.

The white colour of these cattle is dominant. Their muzzles, hooves and tips of the horns are black. Ears, and spots on the lower parts of the legs can be either black or red.

There is a domesticated breed of cattle very similar to the Wild White in colouring and supposedly derived from it. This is the British White which is polled and a better commercial animal. At one time it was a popular dual purpose breed in the eastern counties of England, but it is now rare.

A Breed Society was formed in 1918, called the Park Cattle Society, which registered both horned and polled beasts. Today, they are known as the British White Park Cattle Society and have an appendix to their Herd Book, which is solely for horned animals. Between 1946 and 1973 only polled animals could be registered.

Longhorn

Neolithic cave paintings at Lascaux in south-west France show cattle with widespread horns and a white line down the back. These characteristics are common to Longhorn cattle and indicate that this breed is descended from the extinct wild auroch of Europe. It is not certain how the breed came to Britain, but one theory is that the first longhorned animals were introduced by the Romans.

In medieval times the cows of this breed were milked, and the steers used to pull ploughs. At the end of their working life they were fattened and driven many miles to the meat market at Smithfield in London. This triple purpose breed became one of the most popular types in Britain prior to 1800. It was known in different parts of the country as the Lancashire, Leicestershire, Warwickshire or Dishley breed.

Many men worked to improve the Longhorn cattle, but the most famous was Robert Bakewell of Dishley Grange, Leicestershire. He borrowed a technique that had previously been used by race-horse owners. It was called inbreeding and involved mating a bull to his daughters, his sisters or his dam. This was considered sinful in the mid-18th century, but after a time became permissible. Robert Bakewell and his contemporaries produced earlier maturing animals with thinner hides and finer bones. Unfortunately at the same time they depressed the milk yields.

Longhorn cattle have a dense, silky coat of long hair that varies in colour from roan to dark red and brindle. In all animals, a white line runs along the back and down the tail. This 'Finch-back' or 'Line-back' as it is sometimes called is common to many of the early breeds. There is also a white patch on each thigh, and the lower parts of the leg and brisket are often white. The long, broad head is either roan or white in colour and carries the characteristic large horns. These can be a 1 m (3 ft) or more from tip to tip and either have a forward sweep or curve downwards and inwards towards the mouth.

There are now very few herds of Longhorns left in Britain as they were replaced by the Shorthorn and other more modern breeds.

Old Gloucestershire

Cattle of this breed are thought to share a common ancestry with the Longhorn and Welsh Black. The animals are very similar to the now extinct Glamorgan breed and nearly followed them into extinction. A Breed Society was formed in 1919 to preserve these attractive beasts, but in 1966 it was closed down. Fortunately, interest in the Old Gloucestershire cattle re-awakened and the Society was started again in 1973.

The animals are a mahogany brown with black head and legs. A white line runs along the back, down the tail, between the hind legs and along the belly to the brisket.

Gloucestershire cattle were at one time very popular in the county of Gloucester as they produced a high quality milk. This was ideal for the local cheese industry, especially as the small fat globules in the milk made for a fine, even-textured cheese. As these fat droplets only rose very slowly to the top of the milk, it had to be skimmed twice. This led to the name of 'Double Gloucester' being given to the famous local cheese.

As well as providing milk, butter and cheese, the Old Gloucestershire breed produced oxen for work purposes. These strong, docile animals with their neat horns, could be yoked together to form a team for ploughing or pulling a cart. In fact, they were so quiet that it is reported that one of the hunting Dukes of Beaufort encouraged his tenants to keep Gloucesters as they were not disturbed by his pack of hounds!

Old Gloucestershire cattle also produce good beef carcases and therefore were triple purpose like most of the ancient domestic cattle.

West Highland

The West Highland, or Kyloe as it is known locally, is the native breed of the Western Highlands and Islands of Scotland. These picturesque animals have been kept in the north-west of Scotland since time immemorial. The breed has never become commercially important away from its native home but has been used by many owners to enhance the scenery around their homes. Many animals were shipped to the U.S.A. for this reason alone.

Hardiness is the outstanding characteristic of this breed. With their long, shaggy coats they can be left outside in the most severe winter without any artificial shelter. West Highland cattle are highly resistant to disease and tuberculosis is unknown in the breed. Cows calve easily by themselves and are renowned for living and breeding regularly to sixteen years of age and more.

Beef from West Highland animals is lean and has an excellent flavour. On rough pastures it can take at least three years to finish them as they are slow maturing. When reared under lowland conditions on an improved diet, they put on flesh more quickly and become useful beef producers.

One particularly attractive feature of the breed is the widespread horns. In the male they are strong and come out level from the side of the head, before turning forwards and slightly up at the tips. Cows have longer horns that tend to turn up sooner.

Various coat colours are allowed in the breed, although red-brown is preferred. Black is now rare but there are brindle, yellow and dun coloured animals.

A Breed Society was formed in 1884 and a year later the first Herd Book was published.

Galloway

Cattle of the Galloway type have been found in the south-west of Scotland for centuries. It is said that the Galloway was never crossed with any other breed and so is the oldest breed of beef cattle in Britain.

The animals are very hardy and can adapt to the most severe conditions of wind, snow, rain and cold. Their heavy hide and long thick waterproof coat protects them. Galloways are good grazers and can exist on coarse grasses that would not support other breeds.

The Galloway is normally black but a dun colour is also found and can be registered in the same Herd Book. Black animals with a wide, white band around, called the Belted Galloway, are a distinct strain of the breed and are registered separately. All are naturally polled but it is not certain where this characteristic came from as many cattle in the Galloway district of Scotland were horned. Writers at the end of the 18th and beginning of the 19th century did mention polled Galloway cattle. Therefore, breeders must have decided that the polled factor was desirable and selected for it. (*Continued on page 47*)

Wild White (England)

Longhorn (England)

Old Gloucestershire (England)

West Highland (Scotland)

Galloway (Scotland)

Aberdeen Angus (Scotland)

Hereford (England)

South Devon (England)

Unfortunately, records concerning the breed were destroyed by a fire at the Highland Agricultural Museum at Edinburgh in 1851.

Originally, the Galloway was recorded in the same Polled Herd Book as the Aberdeen Angus. However, the Galloway Cattle Society of Great Britain, formed in 1878, published their own Herd Book after this date.

Although slightly smaller than the Aberdeen Angus and slower maturing, the Galloway has similar attributes and can thrive in much harsher conditions.

Aberdeen Angus

There is evidence, from carvings of the period, of hornless or polled cattle existing in Scotland in prehistoric times. Sixteenth century records refer to black, hornless animals in the population and these must have been the ancestors of the modern Aberdeen Angus breed.

Old Scottish writings refer to polled animals as 'doddies', 'humble', 'humlies' or 'homyl'. In seeking to improve their herds breeders crossed two similar strains of cattle called Angus doddies and Buchan humlies.

One man who played a large part in shaping the Aberdeen Angus breed was Hugh Watson. In 1808 he took on the tenancy of Keillor in Angus. His father gave him six of the best and blackest cows and a bull. With these animals, and a few that he purchased, Hugh Watson began his breeding programme. His favourite bull was 'Old Jock' who became number one in the Herd Book when it was founded in 1862. A famous cow in his herd was 'Old Granny' who was killed by lightening at 35 years old, having produced 29 calves.

Another great man in the breed was William McCombie of Tillyfour in Aberdeenshire. His Smithfield winning steer of 1867, 'Black Prince', was taken to Windsor Castle for Queen Victoria to inspect.

A number of characteristics have made the Aberdeen Angus internationally acceptable. They are hardy adaptable animals that have been used successfully on many native breeds to improve the beefing qualities. They are early maturing, produce a good ratio of high to low priced joints, and have firm, white fat covering a 'marbled' flesh.

The polled factor is very important as it is dominant. This means that first cross calves out of horned cows are polled. They are easier to handle, need less space for housing and there is no horn damage to hides and meat.

Calves of the Aberdeen Angus breed are born very easily and it is unusual to have to give them assistance. It is for this reason that large numbers of dairy heifers are mated to Aberdeen Angus bulls.

The black, silky hair of the Aberdeen Angus covers a black pigmented skin. Occasionally red calves are produced and a number of herds of this colour have been established.

The Aberdeen Angus Cattle Society was officially formed in 1879 and now there are Breed Societies all over the world.

Hereford

The Hereford breed has taken the name of the county in west-central England where it was developed. Early records describe the animals as large and red with widespread horns. They were originally draught animals that were only eaten when they grew too old to work.

Benjamin Tomkins who began his improvement of the Herefordshire cattle in 1742, is often regarded as the founder of the breed. His son, Benjamin Tomkins the younger, carried on the good work. Over the years they succeeded in producing an earlier maturing type that was shorter in the leg, finer in the bone and easily fattened on the available food and pastures.

It is not certain where the colour pattern of red bodies and white faces came from. However, another father and son team of William and John Hewer are credited with fixing the colour markings by inbreeding. The Hewers exerted a great influence on the breed at the end of the 18th century, for they rented bulls to many other breeders as well as selling stock.

This dominance of the white face of the Hereford means that calves can be instantly recognised even when out of cows of a different breed. This is a very valuable asset to the Hereford breed for purchasers of calves for fattening always know what they are buying.

The Hereford is well known all over the world for its ability to produce meat. The cattle are hardy and good grazers so that they can be fattened on grass. Animals exported to hot climates run the risk of blistering in the sun on the non-pigmented areas. This is why darker bulls are more popular and particularly those with some red colour, or 'brilling' around the eyes for protection.

Down through the years Hereford cattle have consistently been winning major prizes at top agricultural shows. The famous Smithfield Club was started in 1799 and at the Smithfield Show that year a Hereford won first prize.

In 1846 the first Herd Book was published privately and then in 1878 it was taken over by the Hereford Herd Book Society of Great Britain. In 1883 the Herd Book was closed against all cattle whose sires and dams had not been registered. This guaranteed the purity of the breed.

Polled Hereford cattle have been in existence for many years and are gaining in popularity. They were first developed in North America, about 1893, by the use of Aberdeen Angus blood. At a later date Red Poll blood was used and also naturally occurring polls. In Britain, a black Galloway bull provided the poll factor for the British Polled Hereford.

South Devon

Very little is known about the early history of the South Devon breed. It is considered likely that the ancestors of today's animals came to England from the continent of Europe via the Channel Islands. This theory is supported by

blood typing evidence that shows there are common factors relating the South Devon to the German Yellow.

By the beginning of the 19th century the South Devon had become an important breed in the counties of Devon and Cornwall. The cows gave plenty of rich milk from which clotted cream could be made. Steers provided quality carcases and the cattle were quiet and strong enough for work. Now the emphasis within the breed is on beef production but at one time the South Devon occupied an unique position in British agriculture. It qualified for a premium on its milk and also for a beef subsidy on its calves and steers. A quality premium is now only paid on Jersey and Guernsey milk.

The South Devon is the largest British breed and a mature bull was judged to be the biggest in Britain on an estimated weight of 1,678 kg (3,700 lb). The liveweight of a mature cow is 710 kg (1,568 lb). The coat colour is a rich medium red or yellowish red. The head is long and relatively narrow with short horns that have a forward, downward sweep. The neck blends in well with the shoulder, the topline is level, the chest deep and the ribs well sprung. The broad rump is long and level. Overall, the South Devon is a well balanced, deep bodied and evenly fleshed animal.

For many years the breed was only popular in its home counties, but now herds are found all over Britain and in many other countries. One of the reasons for the breed's expansion has been the proof of its fast, economical weight gain through performance and progeny testing.

The meat is well marbled, fine grained and has a good flavour.

Sussex

It is generally accepted that the Sussex breed is descended from the red cattle that lived in the dense forests of the weald land of Sussex and Kent at the time of the Norman Conquest of Britain in 1066. Although recorded at that time, there is no more documentary evidence to say how these cattle were developed until the 18th century.

Originally, the breed was kept mainly for draught purposes and work oxen were to be found in this south-east corner of England long after horses had taken over elsewhere. In the late 1800's teams of eight oxen were still used locally to draw the heavy wooden ploughs, even though steam ploughs were being used in other parts of the country. A team of draught Sussex could be seen working around the village of Ditchling right up to 1940.

The Sussex is now a beef breed which owes its firm muscling, good feet and strong legs to its working background. The coat colour is a deep red with short hair on a fine skin. The horns are fairly large and spreading with a light base and dark tips. Cows do not have very well developed udders but produce enough milk to suckle their calves.

In 1951 a programme was devised to breed a strain of polled Sussex. A red Aberdeen Angus bull 'Red Jake of Cropwell' was used at first and by 1966 a sub-race of polled Sussex cattle was introduced on to the open market. They could be registered in the Polled Section of the Society's Herd Book, which was first published in 1879.

Sussex cattle have been widely exported and adapt well to different climates. They produce good quality beef either pure bred or crossed with

Milk Ro…

Milk cart and horse of 1890

The heavy, iron-tyred 'Pram' of 1924

Ford milk lorry of 1934

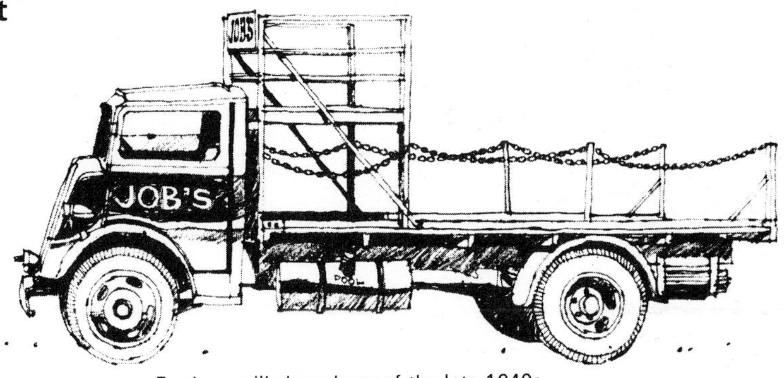

Fordson milkchurn lorry of the late 1940s

Ford van of the 1950s, for milk churns and crates

Standon electric delivery vehicle of the 1960s

AEC Mandator articulated transport for milk crates

native breeds. Recent research has shown a blood factor in the Sussex which is common in *Bos indicus* or 'zebu' type cattle but virtually unknown in other British breeds. This may explain the breed's ability to withstand heat, drought and tick-borne diseases which makes it possible for them to be kept in sub-tropical countries.

Four-year-old males weigh between 870–1,020 kg (1,920–2,250 lb) and females of the same age 560–610 kg (1,235–1,345 lb).

Welsh Black

The Welsh Black is one of Britain's oldest breeds, having been the native cattle of Wales since pre-Roman times. Today's animals are the descendents of cattle which the ancient Britons took with them into the mountains as they retreated from the Saxons.

Different types evolved throughout Wales, but these were amalgamated, until by the time the Breed Society was formed in 1873, there were only two distinct types left. These were the North Wales or 'Anglesey' breed, which was a compact, sturdy, beef strain, and the South Wales or 'Castlemartin' breed, which were bigger, rangier, dual purpose animals. Because of this difference, separate Herd Books were maintained until 1904, when the two were brought together as the Welsh Black Cattle Society. Members are now intermingling the two types.

Although traditionally known as a dual purpose breed, the Welsh Black is now regarded primarily as a beef producer. Welsh Black cows are ideal mothers for their beef calves as they do not produce too much milk just after calving. They provide a steady yield throughout their lactation, so that their calves can be suckled for many months.

Being bred on the Welsh hills for hundreds of years has made this breed very hardy. The cattle can be out-wintered and have a remarkable digestion which allows them to thrive on inferior grazings. They develop a long winter coat that has a dull brown tinge and is shed in spring. If the cold persists, then the longer hair will be retained. If kept in hot climates, the cattle remain sleek.

The head of the Welsh Black is short and broad with long spreading horns that are yellowish white with black tips. The skin and muzzle have a black pigment and the hair is black although small amounts of white are allowed on the underline behind the navel. Cows have moderate to large udders and when mature can weigh 813 kg (1,792 lb). Mature bulls can scale 1,118 kg (2,464 lb).

Lincoln Red

The Viking invaders from Scandinavia landed all along the east coast of England and brought with them the first really large cattle. Descendants of these big, rugged beasts became firmly established in Lincolnshire and the eastern counties of Britain.

These cattle were improved by using Shorthorn bulls, in particular some of the red coloured bulls from the herd of Charles Colling. One breeder who is credited with developing the breed is Thomas Turnell of Wragby in Lincolnshire.

To begin with, these cherry-red, large-framed animals were known as Lincoln Red Shorthorns. They were registered in the first volumes of Coates (Shorthorn) Herd Book. In 1896 the Lincoln Red Shorthorn Society published its own Herd Book.

A memorable year for the breed was 1939 when Mr. E. L. C. Pentecost of Cropwell Butler in Nottinghamshire began work on the development of a polled strain. This was the first time that any British cattle breeder had tried to breed the horns off a traditional horned breed. Red and black Aberdeen Angus bulls were used and after seventeen years the first polled Lincoln Red bull was accepted by the Ministry of Agriculture for licensing. Today, the majority of animals in the breed are polled.

Selective breeding with these originally dual purpose animals produced two distinct types. In 1946 the Breed Society decided to divide the Herd Book into two sections. One section for entries from beef herds and the other for dairy herds. Gradually over the years, the emphasis swung towards beef. In 1960 the word 'Shorthorn' was dropped and the breed became officially known as the Lincoln Red. In the following year, the Society became the first in Britain to set up its own beef recording scheme. All animals were weighed at weaning (200–300 days) and again between 350 and 450 days. This became the basis of a national weight recording scheme as other breeds followed.

Traditionally, the Lincoln Red has produced heavy carcases at three years of age from grass and arable by-products. The demand for smaller joints of meat has changed this system. With the Lincoln Red's rapid growth it can produce cattle for slaughter at a younger age and lower weight.

Devon

The Devon is a native of south-west England where it has been producing beef for many centuries. Originally developed in the counties of Devon, Cornwall and Somerset, it can now be found all over Great Britain.

In 1851, the first Herd Book was published by Colonel John Tanner Davey, but pure breeding had been in existence long before this. During the period between 1793 and 1823, Francis

Quartly of Great Champson, Molland had purchased the best individuals he could find. By continued selection and inbreeding he produced a superior strain that has been admired and sought after through the years.

Devon cattle have coats of a dark cherry red colour which has earned them their nickname of 'Red Rubies'. Their skin is an orange-yellow and this pigmentation is very conspicuous around the eyes and muzzle. The horns of the male are thickset and stick out straight from the head. In the female, they are more slender and sweep upwards and outwards.

A poll register was opened in 1961, following the development of poll strains, using Red Aberdeen Angus, Red Galloway, Red Poll and North American Poll Devons.

The breeders of Devon cattle have produced an animal that is early maturing and finishes readily off grass. They are hardy enough to withstand the rough weather that sweeps off the Bristol Channel, docile and easy calving.

As information on conformation and production are considered essential, the majority of breeders take part in the Breed Improvement Scheme and Genetical Survey. This scheme was launched in 1967 and cattle were classified on a points system for various aspects of conformation. All serious faults were also recorded. Modifications were made in 1971 when a system of five grades was introduced. Animals were graded from A down to E and those in the top two grades qualified for a Register of Merit providing their weight gains were satisfactory.

Exports of Devons have gone on ever since the early 17th century, when the Pilgrim Fathers sailed with red cattle from Plymouth to America. They were imported into Tasmania in the early 1800's and into other Australian states from 1826 onwards. South Africa formed a Breed Society in 1917 and two years later Brazil imported its first Devon cattle.

Ayrshire

The home of this breed is the county of Ayr on the south-west coast of Scotland. Quite a number of different breeds were used to improve the poor native cattle of this area from about 1750 onwards. Shorthorn, Channel Island and West Highland blood was used by the Scottish farmers, who must be given credit for trying out these already improved breeds on their cattle.

At first, the breed was called the Dunlop, after John Dunlop who was probably the greatest of the early breeders. Then it became the Cunningham as it spread throughout the district of the same name. Finally, in 1814, the breed was officially recognised as the Ayrshire and in 1877 the Herd Book Society was formed.

Before 1800, many of the cattle of Ayr were black, although browns and mottled colours had begun to appear. Now the preferred colour is red, or brown, with a varying amount of white, but black with white is accepted. The Ayrshire horns are rather large, set fairly wide apart and have a characteristic upward curve.

Developed in a rugged, high rainfall area the Ayrshire is a hardy animal. It is comparatively small and the neat females have level top lines and tidy udders that wear well. They have won numerous prizes over the years in

Sussex (England)

Welsh Black (Wales)

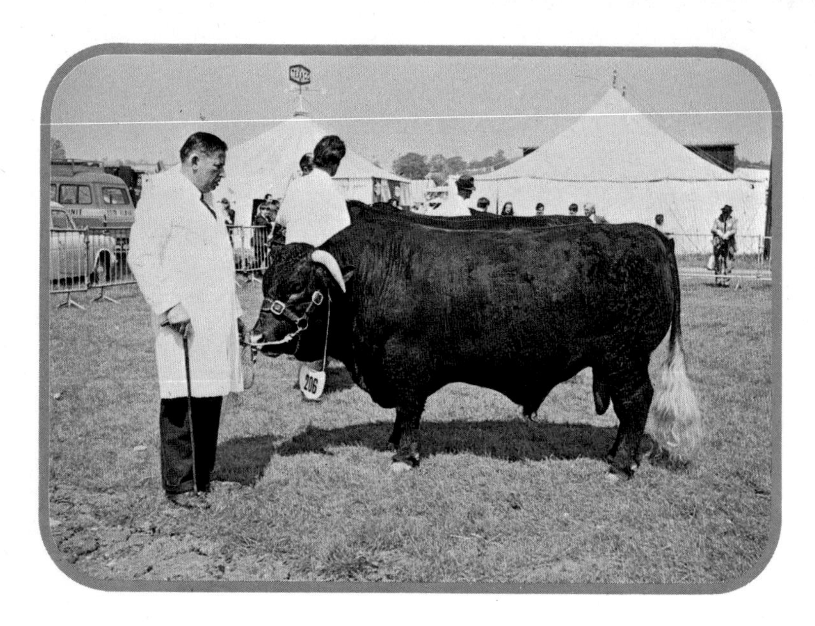

Devon (England)

Lincoln Red (England)

Ayrshire (Scotland)

Shorthorn (England)

Jersey (Jersey)

Guernsey (Guernsey)

agricultural show rings for their perfect dairy conformation. However this has a financial disadvantage in that their bull calves are too fine to be reared successfully for beef unless the cows are mated with beef bulls.

The quantity of milk produced by Ayrshire cows is good in relation to their size. The butterfat content averages about 4% and the fat globules are small, so that they rise to the top slowly. This is an advantage for cheesemaking and so Ayrshire milk is ideal for this purpose.

Shorthorn

As early as 1580 there were superior shorthorned cattle on the Yorkshire estates of the Earls and Dukes of Northumberland. These Teeswater cattle were probably descended from a mixture of the old black Celtic, the red Anglo-Saxon and the broken coloured Dutch cattle. In this north-east corner of England, many breeders were involved in trying to improve their cattle. Between 1730 and 1780 some progress was made and then Charles and Robert Colling became involved. These brothers, living in the county of Durham, really set about developing their animals by means of a selective breeding programme. This was based on the ideas of Robert Bakewell who had successfully improved Longhorns and Leicester sheep.

Two animals bred by the Collings were famous throughout Great Britain. One was the 'Durham Ox', a steer that was exhibited many times, and at five years old was supposed to have weighed 1,371 kg (3,024 lb)! Robert Colling reared a heifer that was twin to a bull and therefore couldn't breed. This animal was fattened for show purposes and became known as 'The White Heifer that Travelled'. Both beasts did a lot to promote the Shorthorn breed.

In 1790 Thomas Booth from Yorkshire purchased his foundation stock and produced a beef strain of Shorthorn. The Beef Shorthorn breed of Scotland was developed from animals imported there in 1837 from the Booth herd.

Another prominent breeder of the time, Thomas Bates, was famous for concentrating on the milking qualities of his cattle. Many of today's Dairy Shorthorns are descended from his stock.

The first herd book established for any breed of cattle was the Coates Herd Book for the Shorthorns. This was a private enterprise started by George Coates in 1822 and taken over by the Breed Society in 1874.

Shorthorns vary in colour from a bright red to a pure white with a mixture of roans where these two colours blend. Horns are short and curve forward in the male and forward and inward in the female.

Numerically, the Shorthorn was once very strong in Great Britain and was exported all over the world. This breed was popular in both North and South America and in Australia a variety called the Illawarra Shorthorn was formed. Now the breed has declined in the face of competition from the specialist beef and dairy breeds.

Jersey

Developed on the island of Jersey, there are now about 8,000 animals on its relatively small surface area. The soil on the island is variable with many rocky outcrops. The farms are small and the average herd size is only twelve animals. In order to make maximum use of the grass available the cattle are tethered to stakes by short chains and moved on every few hours.

Although the origin of the breed is unknown, the best evidence suggests that their early ancestors came from France. Jersey Island is only 15 miles from the nearest French coast and cattle resembling those on the island can be found in Normandy and Brittany.

The Jersey breed on the island has been kept pure for many generations by means of stringent import regulations. The first of these regulations was passed in 1763. In 1833 the Jersey Agricultural and Horticultural Society was established to improve cattle breeding and general agriculture. The following year this Society received the royal patronage of King William IV and has had royal patrons ever since.

In fact, the present monarch, Queen Elizabeth II has a pedigree Jersey herd on her farm at Windsor in Berkshire, England.

The Jersey Herd Book was founded in 1866 but even before this there was a great deal of attention paid to improving the breed. Milk recording was introduced to the island in 1912 and this provided breeders with a valuable guide to the production capabilities of their cattle.

Now the Jersey breed is established world-wide and many factors contribute to its popularity. The animals adapt well to all climates, mature earlier than many other breeds and live long productive lives. Jersey Island breeders stress that a good cow must be able to produce ten times her own body weight in milk annually, with about 5% of fat.

The fine skin of the Jersey is covered with soft hair. Coat colour varies, from fawn to brown and brownish black, either whole or broken up with white patches. Heads are short and have a characteristic concave, 'dish' between the eyes. The dark muzzle is surrounded by a light coloured ring of hair. These elegant, fine-boned animals, with their very good udder conformation, combine beauty and refined dairy type.

AVERAGE BODY DIMENSIONS OF MATURE ANIMALS

Weight

Males—680 kg (1,500 lb)
Females—390 kg (860 lb)

Height at Withers

Males—125 cm (49 in)
Females—120 cm (47 in)

Guernsey

The original home of the Guernsey breed is the small island of Guernsey in the English Channel. It is approximately 9 miles long by 5 miles wide and lies about 30 miles from the French coast.

Robert, Duke of Normandy, sent a colony of monks to the island of Guernsey about A.D. 960. Because these monks lived mainly on a diet of milk, butter and cheese they took cattle with them from France. The animals were thought to have been Froment du Léon cattle—now a rare breed in France. These brown and white cattle, although small, produced a rich milk from which excellent butter could be made.

About 1060, other monks arrived on the island from the Norman city of Cherbourg. They brought with them a large animal known as the Isigny. The crossing of these two breeds produced the new breed that is established throughout the world as the Guernsey, after its place of origin.

By the beginning of the 19th century, Guernsey cattle were famous for the production of large quantities of high quality milk. An export trade built up and in order to keep their breed pure the Royal Court of Guernsey passed an ordinance in 1824 restricting the importation of cattle. In 1878 the first Herd Book was published.

The yellowish skin of the Guernsey is covered with fine hair. The coat is usually a shade of fawn, with or without white markings. The muzzle is cream coloured, the hooves are amber and the tail switch is white. Guernsey cows have long, narrow heads, large gentle eyes and wide nostrils.

Breeders of Guernseys have paid considerable attention to the high quality and colour of the milk which is a very attractive golden yellow. In England 'Golden Guernseys' as they are known produce about 3,600 kg (7,940 lb) milk at 4.6% fat.

AVERAGE BODY DIMENSIONS OF MATURE ANIMALS

Weight

Males—750 kg (1,653 lb)

Females—500 kg (1,102 lb)

Height at Withers

Males—142 cm (56 in)

Females—125 cm (49 in)

Kerry and Dexter

Kerry cattle are the original native cattle of Ireland and have existed as a distinct type for over 3,000 years. This breed takes its name from the County of Kerry and was once very numerous in the south and west of Ireland. Until the 19th century the Kerry was the only breed in Ireland, but now it has been replaced to a large extent by more modern breeds.

The soft, glossy black coat of the Kerry covers a dark skin. A small amount of white on the udder and underline does not disqualify animals from the Herd Book. The medium size head has thin, tapering horns which turn upwards, curve inwards and then backwards at the tips.

Although one of the smallest of the dairy breeds, the Kerry has good body capacity for its size. These active, hardy animals are good foragers; able to produce reasonable milk yields from the poor hill pastures. The Kerry has been known for years as 'the poor man's cow' because of its ability to thrive where other cattle would starve.

Mature Kerry bulls, like the one illustrated on page 166, can reach 550 kg

(1,212 lb) and mature cows average 375 kg (827 lb) and stand 122 cm (48 in) at the withers.

The Dexter is an interesting breed that has been developed from the Kerry; unfortunately it is now extremely rare. In the late 18th century, Mr Dexter who was agent to Lord Hawarden in County Kerry, wanted to produce small cattle that were suitable for beef and milk. He mated a Kerry bull with a very deep bodied cow with short legs and a large udder. The offspring were the foundation stock for the Dexter breed.

Dexter cattle are either all black or all red and are very neat and compact. The short legs and deep body give a dwarfish appearance. Cows milk well, and the meat produced from purebred steers is of good quality. At one time there was an export demand for these small cattle because of their hardiness and heat tolerance, but now the breed is declining rapidly as larger cattle are preferred.

Friesian

Friesian cattle are found all over the world and can be divided into three main types. The Dutch Friesian was developed in Holland which is the home of the breed. The Holstein-Friesian is the North American variety and the British Friesian is intermediate between the other two types.

Unfortunately, the origin of the breed is not at all certain. There is a traditional view that the early ancestors of todays Friesians were domesticated over 2,000 years ago in the flat, marshy lands of North Holland and West Friesland. Another theory is that Friesian and Batavian tribesmen from central Europe introduced to Holland high-yielding cattle derived from ancient Greek stock.

Not many cattle from this early population survived the various disasters that occurred in Holland. There was great loss of stock from both serious floodings and outbreaks of disease. In order to maintain the great cattle industry that had built up in Holland since the 13th century, many animals had to be imported.

BLACK AND WHITE DUTCH FRIESIAN

Dutch cattle were not divided into breeds until late in the 19th century, when Herdbook Societies were established that are still in operation today. The Netherlands Cattle Herdbook Society was formed in 1874 and now has a section for each of the three national breeds.

In 1879, the Friesian Herdbook Society was formed and so there are two Societies that register Friesian cattle in Holland. However, both Societies work closely together and all black and white cattle that conform to the recognised standards are considered to belong to the same breed.

In the early days of the breed, Friesian cattle were selected primarily for their milk production coupled with longevity. Since 1900, breeders have also concentrated on improving the butterfat content of the milk. Later still, selection was for improved beef qualities.

The black and white colour pattern of the Dutch Friesian is varied but the animals must have white tail switches and white on the lower parts of the legs.

A typical cow of the 1950's would be deep bodied, on short legs and with a broad, level rump. The udder was not particularly well attached to the body, with forequarters less well developed

British Friesian Cow (Great Britain)

British Friesian Bull (Great Britain)

than the hindquarters. This 'tilted' udder very possibly had large teats. Today, the conformation of the Dutch Friesian is much improved. The cattle are taller and the neater, level udders are better attached.

HOLSTEIN-FRIESIAN

The early Dutch settlers in North America took their cattle with them and there are references to black and white cattle on the American continent as early as 1621. These animals were mixed with other breeds as were later importations.

The first purebred Holstein-Friesian herd in North America was established by Winthrop W. Chenery of Belmont, Massachusetts. In 1852 he bought a cow that had come from Holland on a Dutch ship. This animal had provided fresh milk for the sailors on the voyage and was such a good cow that Mr Chenery decided to import some more.

In 1881, the Holstein-Friesian was imported into Canada from the U.S.A. and two animals from this original group were to become the ancestors of most Canadian Holsteins.

Holstein-Friesians are famous for their exceptional milk production but the Americans also concentrated on type. A committee of outstanding dairymen was set up in 1922 to consider what should be the correct type of Holstein cattle. Models and paintings were created by artists under the supervision of the committee and widely circulated. A Herd Classification Programme was started in 1929 and breeders' cows were marked on a score card by an official inspector. Owners could compare their cows against the ideal animal and strive to breed better looking beasts.

Holstein-Friesian cattle are the largest of the dairy breeds. They are not as early maturing as other breeds but develop into tall, deep-bodied animals with well attached udders.

BRITISH FRIESIAN

In the 17th and 18th centuries, there were large importations of black and white Friesian type cattle to Britain from Holland. These animals and their descendents produced plenty of milk, and many were bought for the old urban dairies that provided milk to the expanding population of the towns. The Dutch cattle were crossed with local animals, but no real improvement was made until the second half of the 19th century. At that time individual breeders in England started keeping records of the milk yields of their cows. In this way, the best animals could be identified and used to breed the next generation.

The British Friesian Cattle Society was founded in 1909 and a year later issued its first Herd Book. The early members soon sponsored an importation of Dutch Friesians from Holland, and just a few days before the outbreak

AVERAGE BODY DIMENSIONS OF MATURE ANIMALS

Friesian Type	Weight kg (lb)	Height at Withers cm (in)
Dutch male	950 (2,095)	145 (57)
Dutch female	650 (1,433)	135 (53)
Holstein male	1,000 (2,205)	152 (60)
Holstein female	680 (1,500)	145 (57)
British male	1,000 (2,205)	147 (58)
British female	650 (1,433)	140 (55)

of World War I, the cattle were delivered. This importation had a great effect on the new breed as many bulls bred from these Dutch cattle became herd sires. Other importations followed from South Africa in 1922, Holland again in 1936 and 1950, and Canada in 1946. These new bloodlines were blended with the existing stock and in the process the conformation was greatly improved and the butterfat percentage of this high yielding breed was raised. This has only been possible because the British Friesian Cattle Society maintains an 'open' Herd Book, which allows members to register Black and White cattle that conform to set standards. As a result, members have been able to utilise the best bloodlines from anywhere in the world in order to develop their breed.

The British Friesian is known primarily as a dairy breed, yet it also contributes greatly to beef production in Britain as 60% of the meat comes from the national dairy herd which is 80% Friesian. Pedigree and non-pedigree Friesian calves are reared for beef as well as crossbred calves by beef bulls out of Friesian cows. Barren cows and those culled for poor milk production go to the butcher, together with bulls that are no longer required for breeding.

In the last 25 years, the British Friesian has advanced to the position of the dominant dairy breed of the United Kingdom. Over 90% of the total milk supply is provided by this one breed and in officially milk recorded herds the average 305 day lactation is 4,950 kg.

RED AND WHITE FRIESIAN

A number of Black and White Friesians carry a red recessive factor. If two such animals are mated, there is a one in four chance of a Red and White animal being born. Some countries have started Breed Societies for these cattle as they perform as well as their Black and White relatives.

Normandy

The Viking invaders of the 9th and 10th centuries brought cattle with them to France. It is very likely that the Normandy breed is descended from these animals. Originally found in the departments of Manche and Calvados to the south-west of Le Havre, the breed has now spread over most of north-west and central France.

The first attempts at improvement in type and milk production were recorded in the 17th century. In the 18th century, neighbouring areas were using these better animals to improve their own stock. Around 1850, Shorthorn blood was imported from Britain to give early maturity and better conformation.

In 1883 the Herd Book was created so that the work of the breeders could be co-ordinated. Selection was for a dual purpose breed for milk and beef. The Herd Book was re-organised in 1920 and in 1946 a very important decision was taken to milk record every cow registered.

Normandy cattle have a coat colour of dark red-brown, in streaks or patches, on white. Some individuals may have almost black markings and there can also be a yellowish tint present. Under the darker colours, the skin is pigmented and there are patches around the bulging eyes which protects them from the sun.

The Normandy breed has a very good reputation for producing high butterfat and high protein milk. In pedigree herds there are many cows giving over 5,000 kg (11,023 lb) milk at about 4·3% fat and 3·5% protein. This quality milk is used to produce the famous French cheeses, such as Camembert and Neufchatel.

Normandy cattle can adapt to all types of beef production as their good fattening ability enables them to be reared on grass or indoors. Whether killed as veal, steer or bull beef, or aged cow, the meat is of consistently high quality.

During the last 100 years the Normandy breed has increased in numbers dramatically. There are 5,500,000 in France today, which is approximately one quarter of the total cattle population.

AVERAGE BODY DIMENSIONS OF MATURE ANIMALS

Weight

Males—1,000 kg (2,205 lb)

Females—650 kg (1,433 lb)

Height at Withers

Males—150 cm (59 in)

Females—140 cm (55 in)

Tarentaise

This breed is descended from an ancient race of cattle kept in the Alps.

Since 1863, the breed has been known as Tarentaise after the name of a valley in the department of Savoie. A Herd Book was started in 1888, and from that time the breed spread over the whole of south-east France. Animals can be found on the clays of the plains at 200 m right up to the poor, stony Alpine ridges at 2,500 m.

Tarentaise cattle are small when kept in the mountains but they do grow to a larger size when living all their lives at lower altitudes.

The coat is an even colour of yellowish fawn, with the females lighter than the males. However, some individuals carry a badger grey colouring which is similar to the markings of their ancient ancestors. Breeders do not like this, neither do they like a bright red or mahogany coat which indicates that some crossbreeding with a red breed has taken place. The muzzle is black and there are black hairs on the ears and on the tail.

Cows of this breed milk well for their size and the milk is rich in fat and protein which means it is suitable for cheese making.

In order to improve yields of pedigree animals there are quite severe standards imposed before an animal can be registered in the Herd Book. Productivity 'stars' are awarded to good conformation cows that have calved regularly and produced satisfactory amounts of milk.

Numerically, this breed is of minor importance in France, but in its particular area with a low cattle population its presence is significant.

AVERAGE BODY DIMENSIONS OF MATURE ANIMALS

(under lowland conditions)

Weight

Males—800 kg (1,763 lb)

Females—540 g (1,191 lb)

Height at Withers

Males—141 cm (55 in)

Females—130 cm (51 in)

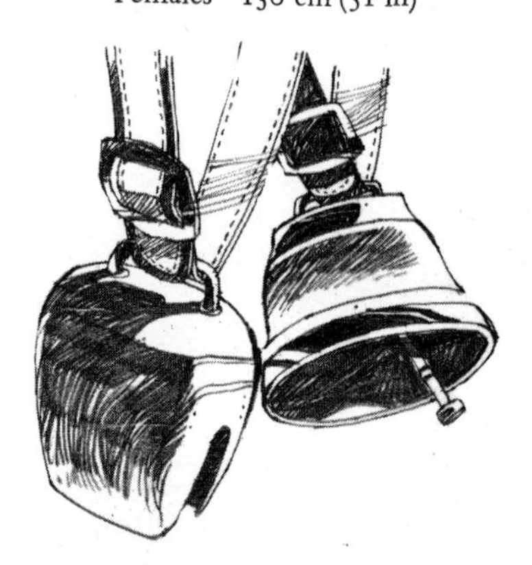

Dutch Friesian (Netherlands)

Holstein Friesian (U.S.A., Canada)

Normandy (France)

Tarentaise (France)

Salers (France)

Aubrac (France)

Charolais (France)

Limousin (France)

Salers

The Salers is one of the oldest of the French cattle breeds. Its name comes from the Salers district, near Aurillac, in the department of Cantal. This is an area of volcanic soils which are acid and poor in mineral content. The valleys, high plateaux, and mountains rise up to 1,800 metres. In this environment, the Salers cattle are sufficiently hardy to endure the bad weather and temperature variations. They can be wintered outside and are used to walking great distances in search of food and water on the most rough terrain.

At the beginning of the 19th century, the master breeder Tyssandier d'Escous saw the potential of this breed and encouraged his fellow breeders to improve conformation. The Salers cattle obtained recognition as a breed in 1853 and a Herd Book was formed in 1908.

The thick skin of the Salers cattle is covered by a medium-to-long, slightly curly hair which is a uniform mahogany red. Those parts not covered with hair, such as the narrow muzzle, are a rosy colour. Calves have brownish horns but the colour changes to ivory with dark tips as the animals grow older. These cattle have large bodies with deep chests and seen from the side look rectangular as the top and bottom lines are approximately parallel. The hindquarters slope down to the tailhead but are well muscled all the way to the hocks. Legs are strong and hooves are hard.

Although the udders on Salers cows are not very well developed, they are capable of producing between 3,000 and 3,500 kg (6,613–7,716 lb) of milk at 3·8% fat in a lactation. This milk has been used in the past for the preparation of Cantal cheeses. Today, the breed is being used more as the basis for the suckler herd. Calves can be left to suckle their mothers for the first nine months, before being sold to beef producers for fattening. Reared in this manner, the calves can achieve an average daily liveweight gain of 1·1 kg (2·42 lb). Very often cows mother two calves, and because of its milk production and growth rates, this breed can supply some of the best French suckling cows.

AVERAGE BODY DIMENSIONS OF MATURE ANIMALS

Weight

Males—950 kg (2,094 lb)
Females—650 kg (1,433 lb)

Height at Withers

Males—149 cm (59 in)
Females—134 cm (53 in)

Aubrac

The home of this breed is southern France in the department of Aveyron. Although no one knows for certain how the breed began, the early development is credited to the monks of the Abbey of Aubrac. This Abbey was destroyed during the French Revolution and the steady improvement of their cattle was interrupted.

There is evidence that the breed was influenced by being exhibited at the

Laguiole Agricultural Show from 1840 to 1880. However, no real pedigree selection began until the Herd Book was started in 1892.

The coat colour of Aubrac cattle varies from light yellow to brown, with darker patches on the shoulders and rump. These darker areas are more apparent on the males. The muzzle, tail switch and hairs around the eyes and in the ears are black. However around the muzzle, along the belly and on the lower parts of the legs the coat is lighter.

This breed can be considered as triple purpose for it was originally developed for meat and draught, but now the milk production is being improved. Milk yields are still not good and vary from 1,800 kg (3,968 lb) to 2,000 kg (4,409 lb) in a lactation.

Animals that are reared for beef cannot be fattened on the relatively poor upland pastures in this southern part of France. Therefore, they are kept in yards and fed a better quality food such as hay, rye meal and oilcakes.

Draught animals can be put to work at about thirty months of age when they are even tempered and willing workers. They used to help in the harvesting and hauling of grapes and grains as well as moving timber. Now they are being replaced by tractors.

The Aubrac is one of the minor French breeds whose existence is threatened by the superior beef and dairy breeds present in the country.

AVERAGE BODY DIMENSIONS OF MATURE ANIMALS

Weight

Males—700 kg (1,543 lb)
Females—500 kg (1,102 lb)

Height at Withers

Males—140 cm (55 in)
Females—130 cm (51 in)

Charolais

The Charolais cattle of today are descended from animals moved into France from Italy by the Roman legions. Until the 18th century, these cattle were kept mainly for their meat production. However, in 1773 Claude Mathieu from Oyé in the department of Saône-et-Loire moved his herd of white animals to Anlezy in the department of Nièvre. Monsieur Mathieu used his cattle for draught purposes and many other herd owners copied him thus producing a dual purpose breed for beef and work.

In order to produce an earlier maturing animal Shorthorns from England were imported to cross with the French cattle. Selection was made for growth rate and carcase quality. Only a limited amount of Shorthorn blood was used in the mid 1800's, as the resulting cattle were less hardy, more temperamental and tended to get fat.

In 1864, the Comte de Bouillé started a Herd Book at Nevers for the Nivernaise-Charolaise breed. Another Herd Book began in 1882 at Charolles for the Charolaise breed. The amalgamation of these two into the Charolais Herd Book took place in 1919.

The Charolais breed can now be found in a wide area of central France comprising the departments of Nièvre, Saône-et-Loire, Allieu, Cher, Loire, Indre, Creuse, Côte d'Or, Vendée and Yonne. The fertile soils of these regions produce excellent pastures and the cattle are kept on flat or gently sloping farmland up to 1,000 m above sea level.

The head of the Charolais is short,

deep and broad, with dark eyes and pale horns growing out sideways, before turning forward and upward at the tips. The medium thick skin is loose and covered with soft hair of medium length. The colour of the hair is white or cream and no other colours are permitted. Under the white coat, the skin and mucous membranes of the Charolais breed contain a light brown pigment. This checks the sun's rays and gives the animals some resistance against sunstroke.

Charolais cattle are large and heavy with long, deep bodies and well sprung ribs. The hindquarters are very well fleshed and the limbs are rather long and heavy boned. Hooves are hard enough for field work, but draught animals that were used a lot on roads had to be shod.

Now that there is hardly any need for draught animals, the Charolais has been developed as a meat producer. In France, between 1850 and 1914, a lot of Charolais cattle were fattened indoors. This method then fell from favour for a while but has been developed again recently with the advent of high quality maize silage. Young animals are kept inside and fed on this silage to produce steer beef at 24 to 26 months of age. If not castrated, the young bulls can be finished at 15 months for 'baby beef' by supplementing their diet with corn. The carcases produced are of excellent quality with a high proportion of intramuscular fat called 'marbling'. This is very acceptable and Charolais bulls are widely used for crossing with other breeds to improve their meat quality. Another advantage is the lack of subcutaneous fat which means that animals can be taken to high weights without laying down a covering of fat over the saleable meat.

A big export trade built up as the potential of the Charolais was realised by breeders in other countries. Cattle have been exported all over the world

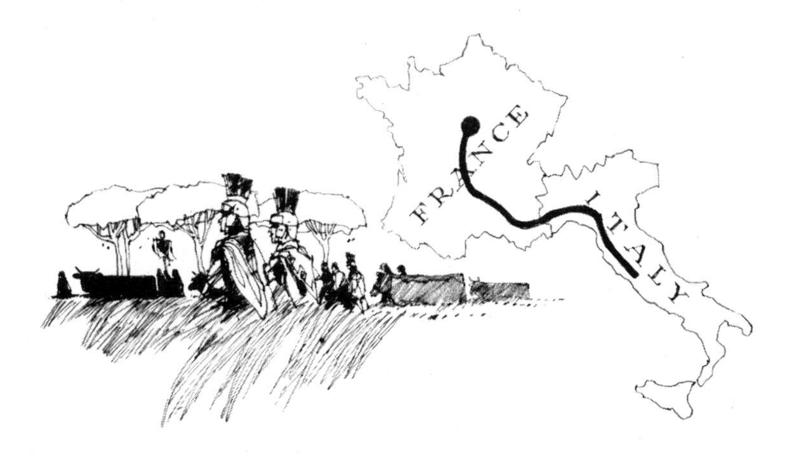

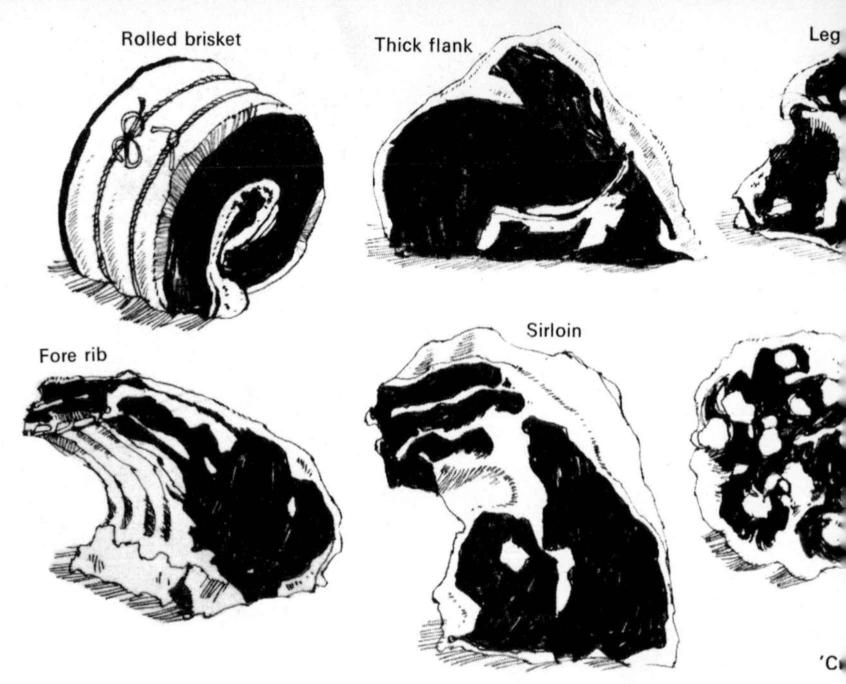

Beef

Cattle can be bred for draught purposes, to provide milk or to produce meat. However, even the cattle that are not specifically reared for beef usually end up as meat when their working life is over.

In most European countries where farm land is scarce, beef production is closely linked to dairying. A large proportion of the meat produced comes from dairy cows, and a still larger percentage from their progeny. Beef producers buy surplus purebred bull calves, as well as crossbred calves by beef bulls out of dairy cows. Some dairy farmers rear their own calves for beef as a subsidiary enterprise.

Less crowded continents than Europe have a different system for supplying beef to their populations. North and South America have large herds of beef cattle, either ranging over vast areas of grazing land or being fattened in intensive feed lots. Not only are their home markets supplied, but they also export to other nations.

Beef, although a popular food, is expensive because beef animals have to be reared for longer than poultry and other meat animals. Cattle are usually kept for at least a year before being sold for meat and many management systems involve feeding them for a longer period. Rearers of beef animals are always looking for more efficient cattle that can consume large quantities of food which they will convert into meat. The quieter animals are more economic as they use up less food to provide energy, so that a larger proportion is used to help them grow.

Cattle pass through different stages on their way to becoming finished

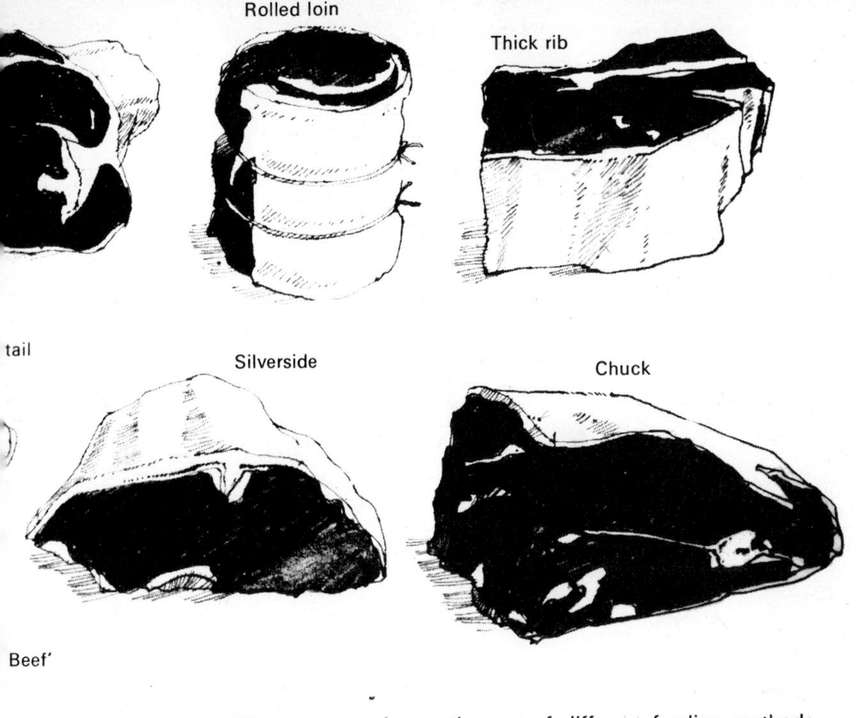

butchers' beasts. When young they have a high proportion of bone and intestines with relatively little muscle tissue (meat). As they grow older the percentage of muscle increases and the animals become longer, wider and deeper in the body. Finally, they reach the stage when the percentage of fat starts to rise. When there is sufficient fat to satisfy the prevailing market requirements, the beasts are slaughtered. Some breeds reach this last stage sooner than others and are called early maturing, e.g. the Aberdeen Angus. Examples of later maturing types that will put on good weights of lean meat before depositing surplus fat are Italian Chianina and British South Devon.

Although there are early and late maturing breeds of cattle it is possible to influence the time taken to finish by the use of different feeding methods. Under poor nutritional conditions the growth of the early maturing types will be stunted and they will remain narrow and leggy. Therefore these breeds are not well suited to harsh conditions. However, given good conditions, the early maturing types are usually more economical to rear as they require less food to maintain them to their earlier finish.

The quality of beef is important to the consumer and here there are distinct differences between breeds. Some breeds store their fat deposits around the internal organs and in a thick layer under the skin. This is wasteful as the butcher has to trim off excess fat before selling the beef.

in order to boost the beef output from native cattle. In 1951 the U.S.A. formed a Breed Society, in 1955 one was started in South Africa and the U.K. followed suit in 1962.

AVERAGE BODY DIMENSIONS OF MATURE ANIMALS

Weight
Males—1,086 kg (2,394 lb)
Females—812 kg (1,790 lb)
Height at Withers
Males—142 cm (56 in)
Females—137 cm (54 in)

Limousin

This breed takes its name from the province of Limousin in west-central France. The area with its poor granitic soil and harsh climate is bordered on the east by the Massif Central and falls from 950 m to 250 m above sea level. Having been developed in this unfavourable situation, the Limousin are hardy cattle. They can live in the open all the year round and withstand both very hot and cold conditions.

Improvement of the breed began in the mid-1800's and with the establishment of the Herd Book in 1886 came certain selection standards designed to produce a better beast. Although originally used for work and meat, the beefing qualities were concentrated on from about 1900.

The coat colour of the Limousin is a dark yellow brown. The head is short, with a broad forehead and broad muzzle. Thin, medium length horns of a blonde colour come out from the head before turning forwards. A shortish neck is followed by smooth shoulders and a long loin. The chest is deep, the ribs are well sprung and the rump is large and well fleshed. The legs are relatively short and strong in spite of being fine boned, and end in hard, blonde hooves.

Outstanding features of this breed are the fine skeleton combined with good muscle development and very little fat covering. French trials have shown that Limousin carcases contain more meat and less bone than those from other breeds.

The ability to produce meat which is marketable at any age is another characteristic of the Limousin breed. Calves can be sold for veal at three months after having been fed on milk only. At the other end of the scale is the 'Châtron'. This is a male calf which is castrated at about ten months and then fattened by three years of age after two summers out on grass. In between these extremes is the 'Saint-Etienne' calf which is suckled by its mother for a while and then finished on hay and concentrates at about ten months. The 'Lyon' calf is reared in a similar way but to thirteen months of age.

AVERAGE BODY DIMENSIONS OF MATURE ANIMALS

Weight
Males—1,100 kg (2,425 lb)
Females—600 kg (1,323 lb)
Height at Withers
Males—140 cm (55 in)
Females—130 cm (51 in)

Blonde d'Aquitaine

Blonde d'Aquitaine cattle are found mainly in an area of south-west France bordering the Pyrenees. This breed was formed in 1961 from an amalgamation of three similar strains of cattle, the Quercy, Garonne and Pyrenean Blond. The Quercy and Garonne strains used imported Beef Shorthorn blood from England to improve conformation.

However, as this led to a less efficient working animal, the Quercy breeders tried crossing with Limousin to re-introduce the draught qualities.

Because of its history, there is a large variation in type within the breed. The French are overcoming this by an intensive breeding programme involving their Artificial Insemination organisations.

Cattle of the Blonde d'Aquitaine breed are wheat-coloured but do vary from almost white to light brown. The legs and belly tend to be a lighter shade than the rest of the body. The medium length heads are narrow and the blonde horns are darker towards the tips.

The animals have deep chests and strongly muscled bodies with wide hips. The strong legs are not coarse and the feet with their blonde coloured hooves are very hard.

The Blonde d'Aquitaine is predominantly a beef breed with little milk potential. Calves are large at birth averaging 47 kg (104 lb) for males and 43 kg (95 lb) for females.

AVERAGE BODY DIMENSIONS OF MATURE ANIMALS

Weight

Males—1,150 kg (2,535 lb)
Females—750 kg (1,653 lb)

Height at Withers

Males—147 cm (58 in)
Females—140 cm (55 in)

Maine Anjou

This French breed was developed in the east of Brittany, between the years 1839 and 1860, by crossing the local Mancelle cattle with the English Shorthorn. The Shorthorn blood improved the conformation, early maturity and milking abilities of the French animals.

In 1908, the Maine Anjou Cattle Society was founded and the members began selecting for increased body size without losing milk production. Maine Anjou cattle have long bodies with thick hairs covering their supple skins. The coat colour is red and white with the minimum of white preferred.

Chests are deep and broad and well muscled shoulders are followed by a straight back. Rumps are long and muscular with the meat being carried right down the thick thighs.

Although a dual purpose breed, milk production is very variable and does not compare with the good beefing qualities. The average annual milk production of an adult cow is only 2,900 kg (6393 lb) at 3·8% fat. However, this relatively low yield can be explained to some extent by the dry summers limiting the growth of grass in the area. Maine Anjou cows do have enough milk to suckle their calves and high growth rates are achieved until weaning.

Reared for beef, these cattle produce an excellent quality meat when fattened on grass or indoors.

Difficult calvings can be a problem with this breed and selection for bulls producing smaller calves is continuing.

The Maine Anjou breed has been exported to many countries and is being used in both purebred and cross-breeding programmes. In July 1974 the World Maine Anjou Council was created in Kansas City, U.S.A. in order to co-ordinate selection and exchange information between breeders.

AVERAGE BODY DIMENSIONS OF MATURE ANIMALS

Weight

Males—1,260 kg (2,778 lb)
Females—910 kg (2,006 lb)

Height at Withers
Males—152 cm (60 in)
Females—142 cm (56 in)

Gasconne

The home of the Gasconne breed is that area of France called Gascony, which extends from the Central Pyrenees to the Garonne basin.

The Gasconne breed is related to the grey 'steppe' cattle that migrated from eastern Europe across Greece, Italy and the south of France. Hungarian Steppe, Greek Steppe and the Italian Romagnola are similar to the Gasconne, as all these breeds have a black pigmented skin covered by grey or white hair. Gasconne cattle have soft, short hair, that is a silver grey, ranging from light to dark.

Two slightly different types of Gasconne animals have evolved, depending on the part of Gascony in which they were bred. On the richer lands to the north of the area, a larger animal with a light coloured coat is found. In the mountain areas there are smaller, darker animals. Both types have medium length horns, which are a yellowish-white near the head and dark at the up-turned tips. Top lines are level, bodies well muscled and the strong legs end in hard hooves.

Before agriculture became mechanised in France, the Gasconne was essentially a draught breed and on the small farms the cows pulled the wagons and ploughs. Calves were reared to three years and then sold to the large farms as work animals. The emphasis has now changed so that the Gasconne is used to produce beef from the mountain pastures. Farmers in the foothills of the Pyrenees try to calve their cows in early March so that the calves can go with their mothers in May to the mountains. One herdsman may look after several hundred animals that are owned by different farmers. This communal herd spends the whole summer grazing the mountain grass.

Three types of beef animals are produced from Gasconne herds. There is the traditional three year old steer that is fed hay during the winter months and then fattened off grass and a little corn. This type of meat production is decreasing and being replaced by the six to eight month uncastrated males that are sold to French or Italian fattening units. These animals go for meat at about 15 months when they weigh about 500 kg (1,102 lb). The third source of meat is the three to four month old calf that is reared entirely on its mother.

Extra income is obtained by the breeders of Gasconne cattle by the sale of females. These hardy animals make good mothers and can rear their calves well. They have a wide pelvic opening and therefore calve easily, which means that they can be mated to bulls of the larger French breeds.

The original Herd Books for the two sorts of Gasconne cattle were started in 1894 and became amalgamated in 1955.

Brown Swiss

Switzerland is divided into 25 regions (cantons) and Brown Swiss cattle are predominant in 18 of these. Found mainly in the eastern half of the country, they are often called the Schwyzer breed after the canton of Schwyz. This was where most of the early work on improving the breed began.

Archaeologists have uncovered skeletons of cattle that indicate the Brown

Blonde d'Aquitaine (France)

Maine Anjou (France)

Brown Swiss (Switzerland)

Simmental (Switzerland)

German Red Pied (Germany)

Angeln (Germany)

Swiss is one of the oldest breeds of cattle. There is evidence that a small red beast was to be found in the lake villages of the area as early as 1,800 B.C. During the last 1,000 years, monastic and other records report the existence of these shorthorned, brown animals. They were kept for meat and work and no improvement occurred until early in the nineteenth century. At this stage in Switzerland's history, land in many of the cantons was fenced and crops of turnips, beets and improved hays were introduced. With this better food supply for the cattle came the desire to breed more productive animals. Improvements in cheese manufacturing were made around 1825 and this created a market for extra milk. Large cattle were brought in from Germany to increase the size.

Today the Brown Swiss is a medium sized animal. The coat is a single coloured greyish-brown that varies in tone, although the darker shades are preferred. There are lighter coloured areas in a ring around the muzzle, in the ears and on the lower parts of the legs. The hair is short, fine and soft and the pigmented skin shows black on the exposed parts such as the muzzle. Horns are white with black tips, medium to small and grow outward, forward and upwards at the tips. The head, with its moderately long face, is wide and slightly dented between the eyes. The back is broad and the topline straight. The chest is deep with well sprung ribs and the well developed hindquarters are nicely fleshed. Legs are rather short and end in dark coloured hooves. Cows have well developed udders with good attachments and neatly placed teats in the better milking strains.

The Brown Swiss cattle are now assuming more of a dual purpose role as milk and meat producers. They only act as draught animals in a few of the more remote areas.

Milk production is good for a dual purpose animal and average yields of 4,000 kg (8,818 lb) at over 3.8% fat can be expected of cows kept under lowland conditions. The beef potential is good with average daily liveweight gains of 1·09 kg (2·4 lb) for bulls and 1·0 kg (2·2 lb) for steers being recorded. The cattle are relatively early maturing and will put on unwanted fat if taken to high weights.

The cattle kept on the plateau and in the lower valleys graze during the summer in pastures between 200 and 800 m above sea level. They have to be housed from mid-October to mid-May but are usually allowed outside for exercise a few hours each day. Young stock sometimes remain outside all year round, only being sent to the Alpine grazings in the summer. At these higher altitudes, between 1,000 and 2,800 m the pastures vary considerably. Animals are kept for the production of calves to sell to the lower farms, or for export.

Thriving at such varying heights, and with the pigmented skin and coat colouring to protect them against the extreme radiation of mountain areas and sub-tropical lowlands means that the breed can live in many countries.

Switzerland's European neighbours quickly saw the advantages of this adaptable breed and Brown Mountain Cattle have been established all over the continent. In France they call them Brune des Alpes, in Italy Bruna Alpina and in Germany they are the Braunvieh, or German Brown.

Brown Swiss cattle can now be found

from the northern provinces of Canada, down through Mexico and Peru where they exist in high lying settlements at around 4,500 m. Many of the Mediterranean countries have Brown Swiss and they can also be found in tropical and equatorial regions from Angola to Central America.

The Breed Society in Switzerland was formed in 1897, a few years after the start of the first Herd Book in 1878. In 1911 the Herd Book was re-organised, and in 1963 the Association of Breeders of Brown Mountain Cattle was formed.

AVERAGE BODY DIMENSIONS OF MATURE ANIMALS

Weight

Males—1,009 kg (2,220 lb)
Females—605 kg (1,330 lb)

Height at Withers

Males—147 cm (58 in)
Females—136 cm (54 in)

Simmental

From its home in the Simme Valley in Switzerland, this breed has spread all over neighbouring countries. It is estimated that there are now more than 35,000,000 Simmental in Europe. The four main types are Swiss, Austrian, German (the Fleckvieh) and French (the Pie-Rouge de l'Est). Other relatively purebred forms of Simmental occur as Czechoslovakian Red Spotted, Hungarian Pied, Hungarian Simmental, Kula (Bulgaria), Polish Simmental, Red Pied Friuli (Italy), Romanian Simmental, Russian Simmental and Yugoslav Pied.

Today, the Simmental breed forms about 50% of the total Swiss cattle stock. Approximately 900,000 purebred animals are distributed west of a line running from the lower part of Lake Constance to Zurich, and roughly south to the Simplon Pass. This area comprises the cultivated plains of north and west Switzerland, as well as the pastures of the Jura and the lower slopes of the Alps in the cantons of Berne, Fribourg, Vaud, and parts of Valais.

Red and White Spotted Simmental Cattle have been present in Switzerland for centuries. They are mentioned in medieval records as being well known for their imposing stature and excellent dairy qualities. A valuable export trade built up as representatives of Central and East European princes came year after year to the valleys of the Sarine, Simme, Kandar and Upper Aar to buy. They took away bulls, dairy cows, beef cattle and draught animals.

The coat colour of Simmental cattle varies through shades of yellow to red, interspersed with white. The head is white, with an occasional coloured spot. Legs and tail are also white and on the body there are white patches, particularly behind the shoulders and on the flanks. The hair is soft, and the skin of medium thickness and lightly pigmented. The cows have well attached udders which, although not large, are capable of high yields of milk.

Originally, the Simmental breed was developed for three purposes—work, beef and milk. The demand for working animals has decreased considerably with the advance of mechanisation in agriculture. However, the characteristics which were necessary for working animals are still important today. To make full use of Alpine pastures during the summer months, the animal must be sturdy, healthy and able to walk with ease. These attributes also result in a long life.

Throughout Central and Eastern Europe, Simmental cattle are the main suppliers of high quality beef. This is either as a pure breed or crossed with other types. Many of the countries outside Europe that have imported Simmental are using them to cross with dairy cows in order to produce calves that can be reared for beef.

The development of the breed has been highly organised over the years from 1890 when the Swiss Red and White Spotted Simmental Cattle Association was formed. Today, the performance and progeny testing of bulls and milk recording of females are an integral part of the breed's progress in Europe. This extensive testing work means that potential buyers have a wealth of data on which to base their selection.

AVERAGE BODY DIMENSIONS OF MATURE ANIMALS

Weight

Males—1,080 kg (2,381 lb)

Females—750 kg (1,653 lb)

Height at Withers

Males—144 cm (57 in)

Females—138 cm (54 in)

German Red Pied

Found over a large area of North-West Germany, the German Red Pied breed has been formed from eight separate breeds early in this century. The breeds losing their separate identities were only formed between 100 and 200 years ago. There was the Breikenburg of Schleswig-Holstein, a dairy breed, and from the same area the dual purpose Geest cattle. Other breeds involved were the East Friesland, South Oldenburg, Stade, Waldeck, Westphalian and Rhineland breeds. The mixture of dairy and dual purpose types has resulted in cattle of a similar build to the Dutch Friesian.

Short, soft hairs cover the medium thick skin of the German Red Pied cattle. The coat colour is red and white in clearly defined areas, with the skin under the red hair being pigmented.

(*Continued on page 88*)

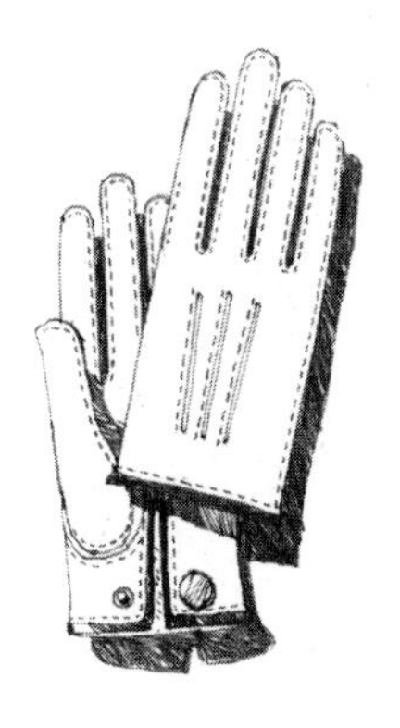

Cattle as Draught Animals

Draught animals or 'beasts of burden' are mostly hoofed mammals. Within this section of the animal kingdom are some of the larger land mammals, and they have shown themselves to be adaptable to working.

Animals were used for draught purposes long before there were any written records. The evidence of this comes from early graves and tomb paintings. In Mesopotamia there were sledges drawn by oxen before 3,500 B.C. and the civilised peoples of the Indus valley had wheeled vehicles drawn by animals as early as 2,500 B.C.

Until 1,000 B.C. only cattle were used for heavy transport. Although they are not as powerful as horses, they can travel better over rough ground and continue to pull long after a horse has given up. In order to increase docility, males are castrated and called oxen, and used instead of bulls. The practise of castration was originally a religious ritual that came to have a practical application.

Early methods of linking oxen to their plough or cart was by a rope tied to the horns. As horns are not always firm and sometimes snap off, other methods were devised. Paired oxen were usually linked by a wooden bar attached to the horns of both animals and then the pulling ropes were attached to the bar. Later still, a beam was laid across the shoulders of the

pair and held in place by thongs passed under the throat of each animal.

Oxen are yoked in pairs for extra power and also for training a young ox by linking it to an experienced animal. This practice of pairing continued through Greek and Roman times. To plough the heavier soils of northern Europe it was necessary to use a larger team of four, or even eight oxen.

In the Middle Ages there was great competition between the oxen and the horse for draught purposes. The swifter horse won in Europe and in many countries has completely replaced the oxen, only to be displaced in their turn by machines. However, in Asia and Africa oxen remain the standard draught team in the unmechanised areas.

Human ingenuity has put cattle to many uses apart from the obvious ones of land cultivation and pulling carts. Waterwheels are turned by oxen in Egypt, so that the irrigation channels continue to flow. Sugar cane crushers in India and the Sudan are worked by cattle. In Corsica, oxen are driven around a threshing floor dragging between them a huge boulder which separates the grain from the straw. Tibetans use Yaks as pack animals at high altitudes. All over the world cattle have played a vital rôle in human progress.

These well built animals have deep bodies with a good spring of rib. They are broad chested and have a good muscle formation on their wide backs and long rumps. Legs are strong and end in hard hooves whilst udders are large and well attached.

This breed fattens well and produces good carcases of marbled meat. It is possible to fatten animals off grass as well as on concentrates indoors. Carcase yield is between 50 and 60% of the liveweight.

A dual purpose breed must also produce milk in acceptable quantities. Recent lactation averages of all recorded cows show 4,149 kg (9,147 lb) of milk being produced with a fat percentage of 3·77.

In the late 1880's Breeders Associations were formed for each of the breeds that were included in the German Red Pied breed. An overall Breed Society, was established in 1922.

AVERAGE BODY DIMENSIONS OF MATURE ANIMALS

Weight

Males—1,000 kg (2,205 lb)
Females—650 kg (1,433 lb)

Height at Withers

Males—139 cm (55 in)
Females—130 cm (51 in)

Angeln

This German breed has been developed from local cattle that have been in the Angeln region for centuries. Situated in the north-east of Schleswig-Holstein the Angeln peninsular juts out into the Baltic Sea. Its clay and sandy clay soils are swept by strong winds and drenched by high rainfall. In spite of this climate, the hardy Angeln cattle are to be found outdoors on pasture from spring to the first frosts of winter.

Angeln cattle have short, fine, glossy hair that varies in colour from brown through red to dark red. This darker shade is preferred in bulls and any white marks on either sex are not desirable, although are tolerated if on the belly. The muzzle and tongue are dark grey to black, and the hooves are strong and black. The loose, thin skin is darkly pigmented. Medium length heads with broad, slightly concave foreheads carry white horns with black tips. These horns are of medium length, growing out sideways from the head with the tips turning inwards.

The breeders of Angeln cattle aim for a high average yield and have set a target of 5,000 kg (11,023 lb) of milk in a lactation. Butterfat is also important and an average of 5% is sought. Many individual cows are achieving such yields.

Although a dairy type of cattle with large, fine skinned udders and well spaced teats, these animals also have deep chests, well sprung ribs and broad backs. This has meant that recently they have been developed more as a dual purpose breed for milk and meat. Carcase quality has been improved by the selection of superior animals from which to breed.

In 1879, the Association of Angeln Cattle Breeders was formed and a Herd Book began in 1885. The Association works closely with the Government who supervise milk recording, the registration of animals and the maintenance of the Herd Book.

As long ago as 1840, Angeln cattle were exported to Denmark to develop the Red Danish breed. In recent times animals have been sold to the Soviet Union, Baltic States, and Turkey in order to improve their local stock.

AVERAGE BODY DIMENSIONS OF MATURE ANIMALS

Weight

Males—850 kg (1,874 lb)
Females—550 kg (1,212 lb)

Height at Withers

Males—140 cm (55 in)
Females—125 cm (49 in)

German Yellow

The German Yellow cattle, or Gelbvieh, are found in the Franconia region of northern Bavaria. The main breeding area is around the towns of Wurzburg, Bamberg and Nuremberg. The Gelbvieh are descended from local cattle that in the late 18th and early 19th century were improved by crossing with Simmental and Brown Swiss. Selection was for an unbroken colour, ability to work, and good growth. However, development at this time was into four separate breeds—called the Glan-Donnersberg, Yellow Franconian, Limburg and the Lahn. As they evolved along similar lines, it was possible in 1920 to amalgamate them into the present German Yellow cattle breed.

On some of the small farms, which are numerous in Bavaria, the breed is still used for field work. As the region is rugged, drier than average and with low winter temperatures, it is a practice to keep the animals housed all year round. This means cutting and carrying grass, lucerne and clover to them during the summer and feeding them hay and fodder beet in the winter.

Although this breed is called the German Yellow, its coat colour varies from cream to reddish-yellow. They have fine, dense hair, which, combined with a pigmented skin, gives good resistance to the sun's rays.

The relatively short head carries medium length horns that curve forward in males and, forward and downward in females. The body is medium long with a good spring of rib and deep chest. The hindquarters are well developed, the legs strong and the feet good. Udders are of good conformation, well developed and firmly attached.

The Gelbvieh Herdbook Association was founded in 1897. Milk recording was made compulsory in 1934 for all cows in members' herds. A system of planned selection for milk yield was started in 1952. Twenty years later all the Gelbvieh cows in Bavaria averaged 3,626 kg (7,995 lb) of milk whilst herdbook cows averaged 3,957 kg (8,723 lb). These are very creditable records for a breed that is also being developed for its beef potential.

AVERAGE BODY DIMENSIONS OF MATURE ANIMALS

Weight

Males—1,040 kg (2,293 lb)
Females—640 kg (1,411 lb)

Height at Withers

Males—143 cm (56 in)
Females—132 cm (52 in)

Pinzgauer

The home of this breed is in the Pinzgau valley in the Austrian province of Salzburg. Several very old types of cattle have been combined to form the present Pinzgauer breed, which has been developing since the mid 19th century. Numbers of breeding stock have been exported to neighbouring countries.

In Austria, the Pinzgauer cattle are found around Salzburg and in the provinces of Steiermark, Tyrol and Karnten. The majority of the breeding animals are in the alpine area of northern Austria surrounding the Gross Glockner mountain. They graze up to 2,500 m in the summer months but live for the rest of the year at between 500 and 1,500 m.

The Pinzgauer breed is well known for its colour pattern. The basic colour varies from light red to very dark chestnut brown. A characteristic white mark stretches from the withers along the top to the rump, down to the hocks and then along the bottom line as far as the brisket. The tail is white, but the head must be coloured to ensure the presence of protective pigmentation in the eyes and on the eyelids.

The soft, medium to medium-long hair on these cattle covers a loose elastic skin. Leather made from this skin is of very good quality and is used for the soles of shoes.

These cattle have been used for work on farms over a very long period and their muscle development and hard hooves equip them for their tasks. They have broad, deep bodies and well sprung ribs and are basically dual purpose animals for meat and draught. In recent years, the demand for milk has prompted a shift in emphasis towards a meat and milk type in the valleys.

Breed Societies have been formed in Austria, Germany and Italy and each country has a Herd Book in which to register pedigree animals.

AVERAGE BODY DIMENSIONS OF MATURE ANIMALS

Weight

Males—912 kg (2,010 lb)
Females—602 kg (1,327 lb)

Height at Withers

Males—139 cm (54 in)
Females—133 cm (52 in)

German Yellow (Germany)

Pinzgaur (Austria)

Chianina (Italy)

Marchigiana (Italy)

Romagnola (Italy)

Piemontese (Italy)

Meuse-Rhine-Ijssel (Netherlands)

Groningen White-headed (Netherlands)

Relative sizes of mature Chianina and Jersey

Chianina

In central-west Italy, in the province of Tuscany, is the Chiana Valley, where the Chianina breed originated. The large, white, shorthorned cattle brought in by the Etruscans and used by the Romans as sacrificial animals were their early ancestors. Present day evidence of this fact can be seen by studying ancient sculptures of cattle. The types of head are very similar to those on todays' animals.

Chianina are to be found in various topographical areas in Italy, stretching from Florence to Rome and from Pisa to Perugia. Four types of the breed exist, of which the largest is the Val di Chiana found on the plains and low hills of Arezzo and Siena. The smallest variety is the Calvana from Florence, which lives in the mountains in poor climatic conditions. It suffers severe winters and very poor grazing. The other two intermediate types are the Perugina from Umbria and the Valdarno of Florence and Pisa. The difference in size of the various types is accentuated by certain feeding systems. Mountain grazings being one extreme and intensive fattening units the other. As well as grass and legume hays, there are some interesting local feed mixtures. Certain concoctions include the boiling of ruminal contents, fresh blood, olive residues after pressing, potato peelings and surpluses from human food manufacture.

The Chianina has short, smooth, white hair with a dark pigment to its thin skin which gives considerable heat resistance. Pure bred calves are tan coloured at birth and only change to white at three to four months. They have a black tail switch, eyelids and hooves.

The head has a straight profile with a slightly long face. Horns on young animals are black but turn yellowish, except at the tips, after two years. The horns are short, spreading outward from the head and curling forward and slightly upward. Other features are the well developed dewlap, high shoulder, long and wide loin and sloping rump.

The most important characteristic of this breed is its tremendous size. After 2,000 years of history it is recognised as the largest domestic cattle in the world. Mature males can achieve a height of 180 cm (71 in) at the withers and the bull 'Donetto' weighed 1,740 kg (3,834 lb) at the Arezzo Show in 1955—a world record for any bull of any breed.

First used by the Romans, and now by modern Italians, for draught purposes it is possible today in Italy to see big Chianina cows pulling a cart or some agricultural implement. Naturally, there has been considerable selection over the years for characteristics essential to a working beast. As a result, the breed is noted for strong, but not coarse, legs and hard hooves. Animals can cover great distances and have a reputation for speed of work and stamina. However, these willing, docile beasts are now being replaced by tractors.

An official breeding programme was started in 1932 under the sponsorship of the Italian Government. Rules were laid down in order to control entry into the Herd Book. Outstanding animals were mated and their progeny weighed regularly until 24 months of age. Animals that grew well were retained in the Herd Book providing they were true to the type laid down in the Breed Standard.

As long ago as 1870, bulls from the Val di Chiana were being sent to other areas in order to improve the local cattle. Today, there is still a good export trade for Chianina to all parts of the world. Their heat tolerance and high disease resistance make them well suited to the hotter regions of the globe. However, they also thrive in temperate climates.

In its native country the breed is losing its dual role of draught and beef, as the Italians are now concentrating only on meat production. Carcase quality is good, remaining lean up to heavy weights with a high proportion of prime cuts.

AVERAGE BODY DIMENSIONS OF MATURE ANIMALS

Weight

Males—1,300 kg (2,865 lb)

Females—800 kg (1,763 lb)

Height at Withers

Males—170 cm (67 in)

Females—160 cm (63 in)

Marchigiana

This breed has as its ancestors the grey-white type of cattle brought into Italy after the 5th century by the Barbarians. After the fall of the Roman Empire, these invaders settled in a hilly area around Ancona on the Adriatic coast. This is the province of Marche from which the breed gets its name.

Development of the breed has taken place only in the last one hundred years. It was improved by Chianina blood from the mid 19th to the early 20th century and by Romagnola blood until 1928. The official herd book was established in 1930 and from 1932 it has been the practise to breed pure.

Marchigiana cattle have spread from the Marche into neighbouring regions in the lower half of Italy. They can now be found in the provinces of Ancona, Macerata, Ascoli Piceno, Abruzzi, Benevento, Frosinone, Lazio and Campania.

The climate within this region is one of hot, dry summers and cold, wet winters. Conditions are not good for grazing animals and the roughages produced on this land are of poor quality. The hard environment and difficult feeding conditions have resulted in hardy, thrifty, muscular and structurally sound animals. They have an outstanding walking ability because of the hilly region and the selection that has taken place for working animals.

Calves are tan at birth and remain so for about two or three months. They then change to the adult colouration of light grey to white. The hair is short and on the tail switch, in the ears and on the eyelashes is dark. The skin has a black pigment and the tongue and muzzle are also black. Horns are medium size and vary in colour from yellow at the base to white in the middle and black at the ends. They come out straight from the head before curving forward in males and upward in the females.

The head is short and wide, and the

AVERAGE BODY DIMENSIONS OF MATURE ANIMALS

Weight

Males—1,100 kg (2,420 lb)
Females—640 kg (1,408 lb)

Height at Withers

Males—158 cm (62 in)
Females—144 cm (57 in)

body of medium length and cylindrical in shape with a level topline. Very muscular shoulders are followed by a long loin and hindquarters. The chest is deep and the ribs well sprung, whilst the legs are short with strong bones.

Although developed for both meat production and work, the attention is now on the former. With their quiet temperament and efficient feed utilisation, they respond well to good feeding. Veal production, from calves fed on milk from their mothers, and supplemented with hay and concentrates is common in the Marche. They do not exceed five months at slaughter and the males average 250 kg (551 lb) with females averaging 210 kg (462 lb). However, this calf meat production is mainly from surplus females as the males are used to produce 'bull beef'. The young bulls grow so well that they can be slaughtered at between 14 and 15 months weighing 550–600 kg (1,212–1,322 lb).

Marchigiana cows are noted for living to a good age and it is not uncommon for them to be still breeding regularly at 10 years old. At the end of their reproductive life they are butchered for 'cow meat'. When frozen for ten days or more this becomes tender and comparable with meat from younger animals.

Romagnola

This ancient Italian breed was formed by crossing native animals with the incoming Podolic cattle of the Barbarian invasion. The modern improvement of this breed began around 1800 and during the period 1850 to 1880 Chianina blood was used. This gave rise to the 'Improved' type which lived on the plain and was larger than the 'Mountain' type. However the 'Mountain' type were better working animals.

The breed is spread over widely differing terrain in the provinces of Bologna, Ravenna and Forli in north-east Italy.

Owing to the mechanisation of farming, the Romagnola has been developed solely into a beef breed. Since the Herd Book was started in 1956, the Breed Society has increased the early maturity of the breed. The aim is a quality, heavy carcase at a young age.

Calves of the Romagnola breed are tan coloured at birth. Adults have short, grey, silky hair which grows longer in colder climates. There are darker areas around the neck, eyes and shoulders. Females are generally lighter in colour than the males. The skin of the Romagnola has a black pigment and tongue, muzzle, tail switch and hooves are also black.

Romagnola have short faces, large muzzles and broad foreheads. Young animals have black horns, changing to yellow at the base with black tips as they age. These horns are of medium length and often described as 'lyre' shaped. They curve outward, upward and forward in males but the tips of the females twist back.

The breed is not as tall as the Chianina, having relatively short legs. Other characteristics are deep chests, muscular shoulders, wide backs and well developed hindquarters.

AVERAGE BODY DIMENSIONS OF MATURE ANIMALS

Weight

Males—1,100 kg (2,424 lb)

Females—640 kg (1,410 lb)

Height at Withers

Males—158 cm (62 in)

Females—144 cm (57 in)

Piemontese

In north-west Italy are found the Piemontese cattle, whose origins go back to the ancient local cattle of this area. Until 1840, a mixture of imported blood resulted in a variety of types. There were the large, yellowish cattle of the plains that had long legs. These were good draught animals and beef producers, but only mediocre milkers. In contrast was the small, red to straw-coloured Demonte variety found in the mountains. In between these extremes of size were the hill cattle that were used for work and only produced enough milk for their calves. Finally, there was a fourth, double muscled variety called Albese.

After 1840, there was an amalgamation of the various types and the Piemontese became established as a breed in the provinces of Turin, Cuneo and Asti.

Young calves are a tan colour until three to four months of age when the females turn light grey. The males are a darker grey mixed with black hairs on the head, neck, shoulders, lower parts of the legs and underparts of the trunk. The muzzle, eyebrows, ears and tail brush are black as is the skin of the body orifices.

As the horns develop they are black. However, by the time the animals are two years old the base is yellow, leaving the tips black. In males the horns come out horizontally from the crown of the head. Females' horns differ by turning forward and upward.

The male head is short and smaller than the female. Both have a good width between the eyes and large muzzles. The breed is characterised by long bodies with straight, wide backs, deep chests and well sprung ribs. The legs are well boned and muscular. Cows have rather large teats on medium sized udders.

The first Herd Book began in 1887, but closed in 1891. Although some development began in 1920, there was no real improvement programme until the Breed Society was formed in 1934. Standards were laid down at this time and have since been revised. This has resulted in superior beef strains with better feed efficiency. The carcases produced have fine-fibred muscles, which are well marbled and tasty. The double muscled Albese variety gives up to 10% greater carcase yields than normal animals. One great disadvantage is the calving problems associated with this strain and the lower milk production of the cows.

In 1947, a start was made on improving milk yields by identifying superior females from which to breed. Eleven years later, in 1958, the Herd Book was reintroduced.

AVERAGE BODY DIMENSIONS OF MATURE ANIMALS

Weight

Males—850 kg (1,874 lb)
Females—570 kg (1,257 lb)

Height at Withers

Males—145 cm (57 in)
Females—136 cm (54 in)

Meuse-Rhine-Ijssel

This red and white Dutch breed is descended from the same ancient cattle as the Friesian. It was developed in the east and south-east of Holland in the provinces of Overijssel, Gelderland, North Brabant and Limburg. The name of the breed comes from the rivers that flow through this area.

Over the years, these animals have been selected for their colour and for milk and meat production. These cattle gained recognition as a breed in 1906, when the Netherlands Herdbook Society started registering them separately. The Meuse-Rhine-Ijssel is now the second largest breed numerically in the Netherlands, and is expanding.

The coat colour is red and white and the pattern of markings is similar to that of the Black and White Dutch Friesian. Also the two breeds are very similar in conformation although slight differences are apparent in height and musculature. The Meuse-Rhine-Ijssel animals are a little shorter in the leg and heavier muscled on the body. They are considered to be more of a dual purpose animal as they have a better quality carcase.

In the Netherlands, veal producers prefer the calves of this breed in their fattening units. The requirement is for young animals that will reach 180 kg (397 lb) by the time they are four months old. Competing for these Meuse-Rhine-Ijssel calves are the rearers of bull beef. It has been found that numbers of bulls can be housed together and turned into beef before they are 18 months. This method of rearing is on the increase in the Netherlands as more farmers find that the bulls grow well and live fairly peacefully together even though they are not dehorned.

Groningen White-headed

Developed in the province of Groningen in the north-west part of the Netherlands this breed shares a common ancestry with the Dutch Friesian. The Groningen White-headed was officially recognised as a separate breed in 1906 by the Netherlands Herdbook Society. At this time it was described as being primarily a beef breed. Today, it is considered as a dual purpose breed with the milking characteristics now superior to the beefing qualities.

The breeding policy adopted has involved the use of line breeding. This has meant that all the Groningen bulls in use today can be traced back through their sires to one bull.

The colour markings of the Groningen White-headed are very distinctive. The hair on the body is black whilst that on the head, brisket and fetlocks is white. Often the animals have pigmented black patches around the eyes. Red and white animals also occur but are in the minority, being less than 10% of the population.

Body conformation is similar to that of the Dutch Friesian, but the Groningen does have superior muscle development, particularly over the hindquarters. Udders, however, are not as good as the Friesians and tend to have less well developed forequarters.

The Groningen White-headed is a minor breed in the Netherlands and its numbers have been decreasing as the Friesian and Meuse-Rhine-Ijssel breeds have expanded. It is found mainly in South Holland and Groningen provinces on flat pasture land which is below sea level.

Central and Upper Belgian

These cattle are also known as Belgian Blues and are a result of crossing Dutch Friesians and English Shorthorns with the native animals in Belgium during the second half of the 19th century. The blue colour that often arises when Friesian and Shorthorn blood is crossed was felt to be preferable to all the other colours in the mixed population. Steady improvements were made after 1890 as breeders began selecting for a dual purpose type for milk and meat. At the beginning of the 20th century, Charolais were used to help the beefing qualities.

In 1919, a Herd Book was begun and now this Central and Upper Belgian breed accounts for about 45% of the total cattle population. This makes it by far the most popular of the five Belgian breeds. It is found throughout the provinces of Luxembourg, Namur and Hainaut, the southern two thirds of Brabant, the south of Limburg and western part of Liège.

The medium length hair of these cattle is blue and white although black and white animals are accepted into the Herd Book. The head is short and wide with a large muzzle and small horns that curve forwards. The body is well muscled, being wide and deep with a level topline. Males are usually between 1,100 and 1,250 kg (2,425–2,756 lb) when mature and stand 148 cm (58 in) at the withers. Mature females weigh from 700 to 800 kg (1,543–1,764 lb) and are about 136 cm (54 in) at the withers.

Cows of this breed are good milkers and on average produce nearly 4,000 kg (8,818 lb) of milk in a 305-day lactation at 3·46% fat.

Calves weigh about 40 kg (88 lb) at birth and when fattened for beef they grow well to produce good quality meat. An adult bull yields 63%, a steer 61% and a fattened cow 56% of its liveweight in the form of carcase meat.

Danish Red

On the Danish islands there was a mixture of cattle that were variable in colour. These animals were steadily improved from the end of the 19th century by crossing with imported stock from the Jutland peninsula. Angeln cattle were used to give the emerging breed its dairy characteristics and the Slesvig Marsh animals contributed the frame, size and colour. The Danish Red was first officially recognised as a breed in 1878 and a Herd Book was opened in 1885.

The colour of the hair is a deep red, with the bulls slightly darker than the cows. Small white markings do sometimes occur and are considered undesirable. The hair is soft, short and sleek and covers a loose, thin skin which is darkly pigmented. A medium length head carries horns that curve forwards and slightly downwards. The muzzle is slate coloured. Mature males weigh about 1,000 kg (2,205 lb) and are 148 cm

(58 in) at the withers. Females reach an average of 650 kg (1,433 lb) and a wither height of 132 cm (52 in).

Selection within the breed has been for high milk yield and a high fat percentage coupled with high growth rate and good muscular development. This means that the Danish Red is a dual purpose breed for milk and meat. By 1960 it accounted for 61% of all the milking cows in Denmark. Now, however, it is losing ground to the Danish Jersey and the Black and White Danish breed.

Danish Red cattle have been exported to many countries but they tend to do best where the feeding and management levels are high. In Denmark, the dairy farmers have a flair for looking after cattle and several cows have reached very high lifetime yields. In 1966, one cow was recorded as having produced 91,114 kg (200,870 lb) of milk in her life, all at 4·7% butterfat.

Danish farmers were the pioneers of milk recording and bull progeny testing. For many years they have carried out tests on the milking ability of all first calf cows at their progeny testing stations. Measurements of the amounts of milk taken from each of the four teats show that there is an even distribution of milk between the fore and rear udder. This means that the Danish Red is an ideal animal for modern milking parlours that depend on a high throughput of animals per man hour.

Swedish Red and White

There were originally three breeds of native cattle in Sweden. At the end of the 18th century, a number of foreign breeds were imported and were crossed with the native cattle to such an extent that two of the Swedish breeds disappeared. In their place came the Red Pied Swedish cattle and the Swedish Ayrshire. Both were the result of using imported Ayrshire blood on local stock. The main difference between them being that the Red Pied cattle were more muscular due to Shorthorn influence. As these two breeds were very similar in looks and performance it was decided to amalgamate them. This was done in 1928 when the Swedish Red and White Breed Society was formed and a new Herd Book started.

The breed is found all over Sweden but occurs in greater numbers in the middle and south of the country. Here the soils vary from heavy clay to peat and the altitude never exceeds 200 m. The grazing season in the south lasts for about six months and the cattle have to be housed from mid-September to early May. In the north, the animals are indoors for even longer periods.

Swedish Red and White cattle are a cherry red colour with small white markings on the tail switch, breast, belly and sometimes on the legs and forehead. They have a loose, medium thick skin that is lightly pigmented. The medium length heads have small horns that curve forwards and inwards.

In body conformation they are more like Shorthorns than Ayrshires and are considered to be a dual purpose breed.

Milk yields are around 4,300 kg (9,470 lb) at 4·1% fat in a 305 day lactation. Cows usually calve for the first time at about 2½ years and then go on to produce four or five lactations.

Young animals can be fattened successfully for meat production, as they have high growth rates. Mature males weigh approximately 950 kg (2,090 lb) and females 550 kg (1,210 lb). Wither heights average 136 cm (54 in) for mature males and 129 cm (51 in) for mature females.

Telemark

The Telemark is the oldest of the native breeds in Norway. It was recognised as a breed in 1856 and a Breed Society was formed in 1895. At one time it was the most numerous breed in Norway, but after 1930 it declined in popularity and after the Second World War fell to the position of a minor breed.

Originally there was great variation in colour and type but selection standards were laid down which eventually turned it into the most uniform of the Norwegian breeds.

The soft, medium length hair is mainly red. However, there are white patches along the back, on the udder, belly, brisket and legs. The straight head is of medium size and has long lyre-shaped horns which curve outward, forward and upwards with the tips turning back. The short, deep body has good capacity and a fairly straight back. The long rump slopes away from the backbone on each side. The overall appearance is of a wedge shaped, fine boned dairy animal.

Telemark cows weigh about 420 kg (926 lb) and stand 121 cm (48 in) at the withers. Their calves are normally about 28 kg (62 lb) at birth. Milk yields are not very high although individuals have achieved 7,500 kg (16,535 lb) of milk in a 365-day lactation at 4·3% fat.

As Norway is a mountainous country large areas away from the coast are permanently under snow. Some of the peaks in the south reach to almost 2,500 m. The breed is found in these inland districts of southern Norway where it lives at various altitudes and on different soil types.

Blacksided Trondheim and Nordland

This Norwegian breed has developed over the centuries from the ancient Scandinavian mountain cattle. The animals are polled and are closely related to the polled cattle in the north of Sweden and Finland. It has been suggested that they are directly descended from neolithic cattle.

At one time the Blacksided Trondheim, or Røros cattle were thought to be a different breed from the similar, but smaller, Nordland cattle. In 1943, these two strains were brought together in the one Herd Book.

The soft hair on these cattle is basically white with black patches superimposed. The area of black varies greatly from animal to animal. On some it completely covers the head and body, so that there is only a white area along the back, on the legs, brisket and udder. Other animals are almost completely white except for a small amount of black around the muzzle and on the ears. Under the black patches the skin is darkly pigmented. Some red and white

Central and Upper Belgian (Belgium)

Danish Red (Denmark)

Swedish Red and White (Sweden)

Telemark (Norway)

Blacksided Trondheim and Nordland (Norway)

Finn Cattle (Finland)

Finnish Ayrshire (Finland)

Kholmogor (Russia)

animals are born occasionally and are allowed to be registered in the Herd Book.

The native cattle of Norway were small and this breed is no exception. Bulls average 420 kg (926 lb) whilst cows average 340 kg (750 lb). Heifers are usually calved for the first time at about two years of age and may go on milking for five lactations. Yields are not high, but more animals can be kept per hectare, thus increasing the output of each farm unit.

In Norway, the winters are long and the cattle are housed for about eight months of the year. From June to September the herds are usually moved to the *saeters*, or mountain pastures.

Although the Blacksided Trondheim and Nordland cattle are typical diary animals, they can be fattened to produce beef.

Finncattle

Finncattle are an amalgamation of three separate breeds and before 1947/48 each had their own Herd Book. There were the brown and white East Finnish, the brown West Finnish and the white North Finnish breed. It is assumed that the ancestors of all three breeds must have been introduced to Finland at some stage in that country's history as archaeologists can find no remains of very early cattle. Since amalgamation, the white variety has almost disappeared and the brown with white on the back, belly, legs and head is not now so numerous. Only the brown type has increased in number and spread over the other areas.

Finncattle are a medium sized dairy breed. The liveweight of mature females varies from 380–480 kg (838–1,058 lb) and males from 600–750 kg (1,323–1,653 lb). The animals do not carry horns and the poll comes up to a peak on top of the short head. They are very hardy, lean and healthy as a result of the harsh conditions prevalent in Finland at the time these cattle were developing.

The average age of animals in Finnish herds is $8\frac{1}{2}$ years and many cows go on producing until they are 17 years. Milk yields vary greatly according to management and individual lactations can be as low as 3,500 kg (7,716 lb) or as high as 8,000 kg (17,637 lb). The percentage of butterfat in the milk is very high and some cows have a recorded level of 7%. The best animals of the breed are capable of producing their own liveweight of butterfat in a year.

In the north of Finland, the grazing period on pastures only lasts for three months, and in the south for one month longer. Because the animals have to be fed indoors for the rest of the year, this makes livestock farming very expensive. Under these conditions, the Finns calve their animals for the first time at two years, which is early, and aim for good milk yields right from the start.

Most of the Finnish herds are too small to justify keeping a bull and before artificial insemination became widespread it was common practise to share a bull. About 15 to 20 herd owners would belong to a co-operative bull-keeping association and share all costs. At one time, there was as many as 2,500 of these associations all over Finland, but now they are to be found only in the east and north.

Finnish Ayrshire

Finnish Ayrshire can be found all the

way from the south of Finland up to the Arctic Circle. It was originally introduced in the 19th century, along with several other foreign breeds. Because of the primitive feeding and management conditions prevelant in Finland at that time, only the Ayrshire survived.

The first Ayrshires were brought to Finland from Mecklenburg in Germany about two years before the first importation from Scotland in 1847. Other cattle came in from Sweden, until the last shipment in 1923.

The Finnish Ayrshire is partly the result of pure breeding and partly graded up from other stock, such as Finncattle, Angeln, Danish Red, Shorthorn, Jersey and Friesian. By repeated backcrossing to the pure Ayrshire, a breed has been developed that is very similar, but somewhat bigger and whiter, than those animals in Scotland. Mature bulls average 890 kg (1,962 lb) and are 140 cm (55 in) high at the withers. The females are 460 kg (1,014 lb) on average and stand 126 cm (50 in) at the withers.

Two factors that helped the progress of the breed were the introduction of milk recording to Finland in 1898 and the formation of the Breed Society in 1901. Systematic breeding based on performance has produced a breed that has one of the highest fat production levels of any in the world. Average yields are 4,700 kg (10,362 lb) of milk at 4·6% fat.

In order to improve the beef characteristics of Finnish dairy breeds, a performance testing programme has been in operation since 1966. The Artificial Insemination Societies test their young bull calves by feeding them *ad lib* concentrates and roughage from 60 to 365 days. About 220 Finnish Ayrshire bulls are tested each year and 30% of these are culled.

Kholmogor

The Kholmogor is one of the oldest breeds in Russia and it evolved in the Archangel region. These cattle are now widely distributed throughout the U.S.S.R. and have been used to improve other local types. Exports have been made to Poland, Finland and other Baltic countries.

Ancient documents show that Kholmogor cattle were sold to Moscow milk producers who obviously saw the value of the Kholmogor as a dairy animal.

Improvement in the breed has been carried out since 1765 by importing Friesian blood from the Netherlands.

The majority of the Kholmogor cattle are black and white, but the colouring can vary from almost all-white to almost all-black. Some red and white animals are born, while others are brown, or even grey, and white. The medium thick skin is pigmented under the black patches.

Heads are leaner and shorter than the Dutch Friesian and carry medium length horns that curve forwards and upwards. Kholmogor cattle have long bodies and long legs. They are typically lean dairy animals with good size udders. Although the emphasis is on milk pro-

duction, some selection has taken place for beefing characteristics in order to provide the butcher with a better carcase.

Average mature weights of males are 925 kg (2,040 lb) and females 600 kg (1,323 lb). The males average 146 cm (57 in) at the withers and the females 132 cm (52 in) when mature.

Red Steppe

The Russian Red Steppe breed, also called the Ukrainian Red, is an amalgamation of other breeds. In the second half of the 19th century, various local types of cattle had been produced in the Ukraine by crossing the Ukrainian Grey with other breeds. Angeln, Red Dane and Shorthorn were amongst the imported breeds used, whilst the Kholmogor was the other Russian breed in the mixture.

Although the different districts of the Ukraine had cattle of varying production and conformation, they were all amalgamated into one breed by 1923 when the first Herd Book was begun. The number of cattle meeting the requirements for registration in the Herd Book in 1939 was 50,000. The war years seriously interrupted the breeding programme and when the Herd Book was re-started in 1942 there were only 2,905 registrations.

The medium length red hair on the Red Steppe cattle covers a darkly pigmented skin. The muzzle and hooves are black. The long straight head carries a pair of medium length yellowish-white horns that curve forward, upwards and inwards. There is considerable individual variation between animals in this breed due to the very mixed ancestry. However, most have long bodies with fine bones that gives them a dairylike appearance. Although a milking breed, the cows only have small udders.

The average liveweight of mature bulls is about 900 kg (1,984 lb) and cows about 520 kg (1,146 lb). An average male will stand about 140 cm (55 in) at the withers and a female 130 cm (51 in).

The Red Steppe is not a good producer of meat, but with efficient feeding a reasonable carcase can be obtained. The amount of saleable meat from the carcase can reach 58% of the liveweight.

Ala Tau

The Ala Tau breed has been developed since 1931 in the Soviet provinces of Kazakhstan and Kirgizia, chiefly in the foothills of the Zailiski Ala Tau mountain range. Most of the early work was carried out at the 'Alamedin' stud farm where the stock consisted of local cattle that were being crossed with Friesian, Simmental and Brown Swiss. These hardy, low-yielding animals were combined with 1,000 Kirgiz cattle transferred to 'Alamedin' from state farms. The best cows were selected and mated to Brown Swiss and Kostroma bulls. These are two related breeds, as the Kostroma was developed by crossing Brown Swiss with native cattle in the U.S.S.R.

The result of this controlled breeding programme was a first generation of animals that gave more milk than the native Kirgiz cows. Milk yields ranged from 1,745-5,084 kg (3,850-11,210 lb) in a 300-day lactation at about 3.81% fat. All low yielding animals were culled and the second generation bred from the better cows showed a further increase in production. *(Cont. page 112)*

Hides to Leather

Hides and skins that have their original fibrous texture intact and have been preserved by tanning are known as leather. Skins come from the smaller animals such as sheep and pigs whilst hides are from the larger horses and cattle.

There are three distinct stages in the leather industry: the raising of cattle to supply the hides; the tanning and processing of the hides into various types of leather; the manufacturing of the leather into consumer articles.

At the cattle rearing stage there are a number of factors that affect the quality of the hide. The heavier types come from steers and the lighter ones from dairy animals. Just as the sex and breed of the animal can determine the thickness and therefore the weight of the hide, so can nutrition and disease factors. Horn damage lowers the quality, as does hot-iron branding and parasites such as the warble fly.

Fresh hides are cured by drying or salting and can then be stored in readiness for tanning. The tanning process was practised in Egypt, China and India over 5,000 years ago and vegetable tanning is one of the earliest manufacturing techniques. Extracts made from nuts, barks and fruits are used and the hides are soaked in solutions of increasing strength. The process takes place in pits, rotating drums or special wooden vats known as 'paddles'. It is a simple but time-consuming process that can take up to 60 days. However there is a more modern technique called chrome tanning that uses chromium salts and can be completed in three to twelve hours. This process requires precise conditions and careful control. Chrome tanning is the preferred method for most of the leathers used for gloves and clothing.

After tanning, the hide becomes leather and then needs to be finished, or 'dressed', according to the type of use it will have. Leather for shoe soles is normally only levelled and rolled, but the uppers use leather that has also been softened, dyed and waterproofed. Thick leather can be split into two or three layers.

Leather is used for a very wide range of products although about 65% of the total output goes into shoes. Garments account for a further 20% and the rest is made into handbags, luggage, upholstery, saddles, harness and many industrial goods such as diaphragms, gaskets, washers and oil seals.

Millions of people are employed in the leather industry and it is one of the most widely dispersed industries in the world. Many developing countries are finding that leather is a valuable export commodity and that processing provides work for their people. Europe is a big importer of leather and takes many low quality hides and semi-processed leather from South America, India and Pakistan. The quality hides for clothing and upholstery are mostly supplied by the Scandinavian countries. Britain exports to other European countries more than half of its current production of 5,350,000 hides from steers, heifers, cows, bulls and calves. Since the hide is normally worth between 5 and 10% of the total value of the animal, it can be seen how the leather and tanning industry benefits the livestock farmer.

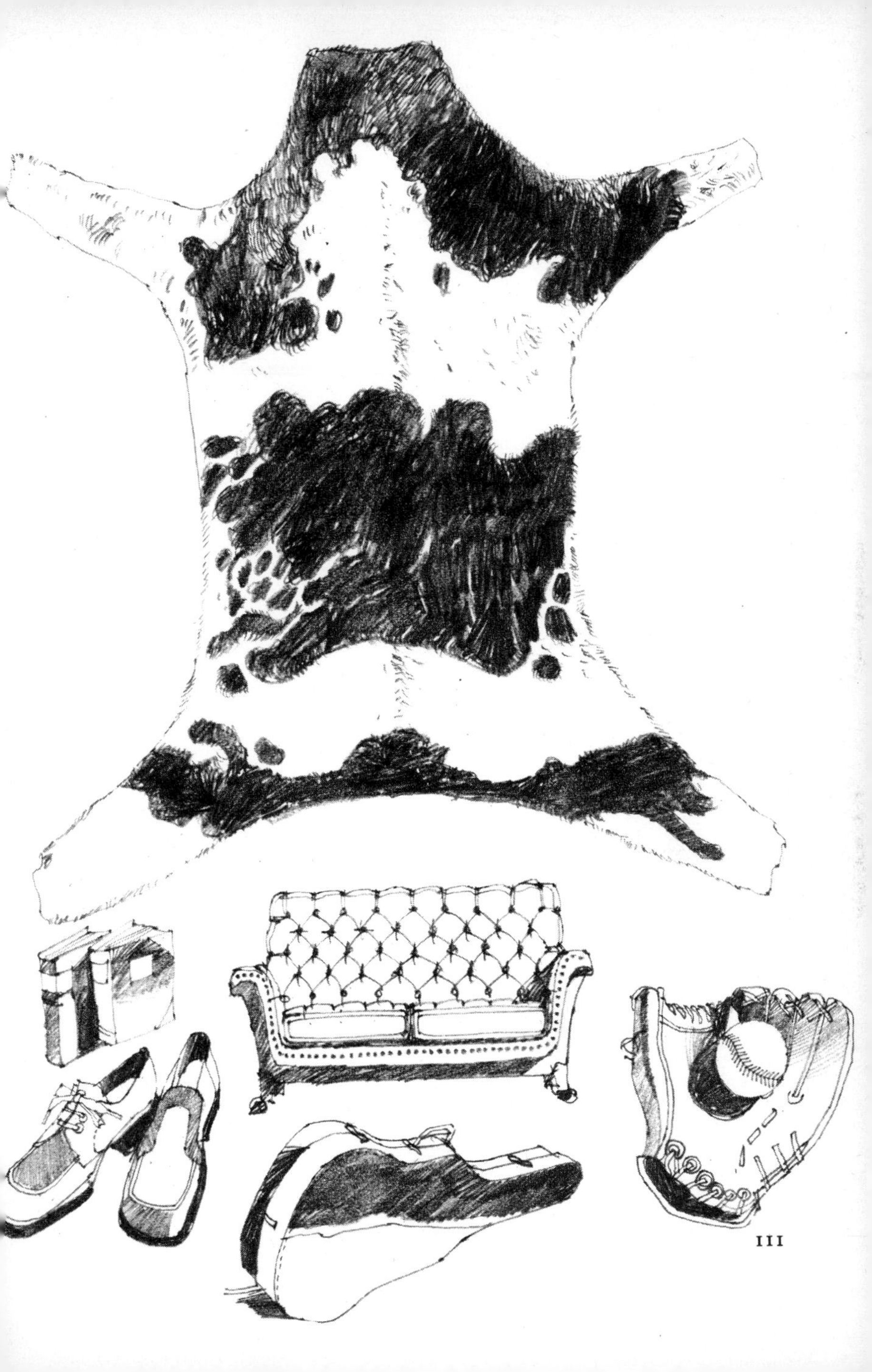

Liveweight increased and good quality carcases were produced; as a result the Ala Tau became dual purpose for milk and meat.

The beef is well marbled, and the saleable meat produced from each carcase is 50–60% of the liveweight, which for adult cows is about 600 kg (1,323 lb).

Ala Tau bulls were carefully selected before being allowed to breed, and occasionally more Brown Swiss or Kostroma bulls were used to introduce fresh blood. After all the pioneering work on the breed had been carried out on the stud farms, the improved animals were used to upgrade the native cattle in Kazakhstan and Kirgizia. There are now some 750,000 pure or grading up Ala Tau in these two provinces alone. These strong, hardy, grey-brown animals have also been introduced to other Soviet States and Mongolia in order to improve their cattle populations.

Monogolian

In Mongolia, the cattle outnumber the people by a ratio of 2:1. This large country of 600,000 square miles is sparsely populated with only 1,300,000 people who share their land with over 2½ million cattle.

The Mongolian cattle are a dual purpose type for milk and meat. They are usually brindle or reddish-brown, but sometimes black or yellow animals occur. Most are small with cows weighing from 280–300 kg (617–660 lb) and the bulls only 360–450 kg (794–992 lb). Conformation is variable and milk yields are low.

Calving takes place in the spring and the cows are milked from May to November, giving an average of 1,500 kg (3,307 lb) during this time. The Mongolian steppe is a level, treeless plain that provides a natural pasture where the cattle are allowed to roam freely. They are controlled by the herdsmen on horseback who look after herds or 'brigades' of 300–400 animals. It is a nomadic life for the herdsmen and milkmaids who follow after the wandering herds and live in wooden huts, or yurts, which they move to each new grazing area. Milking is done by hand with the calves being allowed near their mothers in order to stimulate the flow of milk.

Pneumonia, exposure and starvation cause cattle losses as high as 15% in the very cold winters, even though the animals are moved to sheltered pastures where they are given a little hay. Some herds are kept entirely for beef and the calves are allowed to suckle their dams whilst the older steers are kept separate. These beef animals are butchered at 2½ to 3 years of age and give about 50% of their carcases as saleable meat.

Although at present there are some 750,000 breeding cows of the Mongolian type these numbers are likely to fall rapidly. This is due to recent breeding programmes that have been introduced to improve the cattle in Mongolia. Purebred Friesian, Simmental, Red Steppe and Ala Tau have been imported from Russia to provide a nucleus of breeding stock for the upgrading of the Mongolian cattle.

The Russian Kasakh Whiteheaded, a modified Hereford, is being used to improve the beef herds. Also, semen from British Galloway and Hereford bulls has been tried, as the Mongolians have had an artificial insemination programme in operation for twenty years. Frozen semen has also been imported from the U.S.S.R., Cuba, Czechoslavakia and Japan.

Miranda

Miranda cattle have spread from the mountainous regions of Miranda do Douro and can now be found all over central and north-east Portugal. There are a number of native breeds in Portugal, but none are so numerous or spread over such a wide area as the Miranda. In Portuguese this breed is known as the *Mirandesa*, or sometimes, the *Ratinha*.

The soft, short hair of these animals is dark brown on the males and a lighter shade on the females. The skin is dark-coloured, loose and of medium thickness. Around the muzzle and eyes are lighter coloured patches and the longer hair on the top of the head is also lighter.

The head has a short face with a broad concave forehead and a broad black muzzle. The horns are a medium length, white with black tips. They come straight out from the side of the head before turning downwards and forwards and then out and up at the ends.

The uneven topline of the Miranda is a result of the high withers, low loin and high tail head. Legs are strong boned and muscular, with hard hooves. They are well suited to work and the breed produces the best draught animals in Portugal. Cattle can be put to work at two years of age and will work in the fields as well as on the roads. Although rather slow, they will perform their tasks for eight hours a day if required.

Miranda cattle have not been developed for milk production and are, therefore, dual purpose for work and meat. Young stock are either used for veal production when weaned from their mothers, or for beef when reared to two years of age. Bulls and work oxen are sent to the abattoir from four to seven years and cows usually between seven and ten years. Most stock intended for meat is fed indoors for a period before being sent to the butcher. The carcases produced are normally about 50% of liveweight, or better, depending on age and condition of the animal.

The liveweight of mature males is about 900 kg (1,984 lb) and they stand 143 cm (56 in) on average at the withers. Females are 550 kg (1,212 lb) on average when mature and stand 133 cm (52 in) at the withers.

Barrosa

The Serra do Barroso mountain district in the Tras-os-Montes province of northern Portugal is the original home of this breed. The Barrosa cattle have evolved in this area of low fertility soils and poor mountain pastures. Most of the area is higher than 400 m and some stretches are at an altitude of 700 to 1,000 m.

Although these cattle are bred and

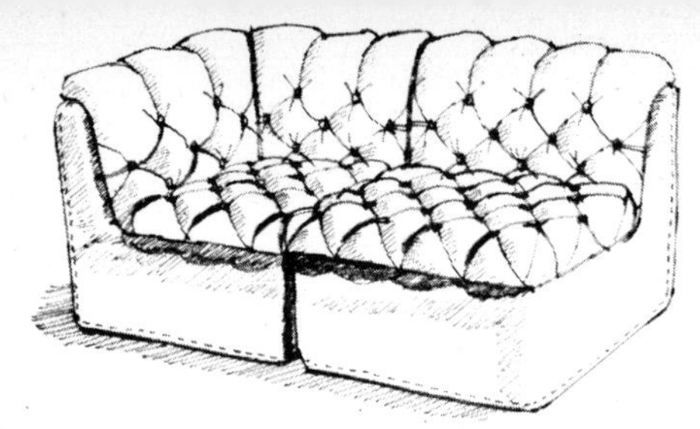

reared in the Serra do Barroso area, they are sold into the Minho province for work and beef production. Minho is at a lower elevation and has a milder climate. A wide range of crops are produced, including maize, and a higher stocking rate can be achieved.

Barrosa cattle have a loose skin of medium thickness which is covered with soft hair. The colour is a reddish-brown with a tendency to become darker at the extremities. The short head has a wide, concave forehead and a broad, black muzzle. Horns are large and lyre shaped, growing outwards before curving upward, inward and backward at the ends. The dewlap is prominent, the chest deep and the ribs well sprung. The short legs are fine boned and end in hard hooves.

Mature males weigh about 580 kg (1,279 lb) and only stand 129 cm (51 in) at the withers. Cows are even shorter at the withers, averaging 122 cm (48 in) and weighing 340 kg (750 lb).

It is possible to work oxen of this breed for eight hours a day pulling a wagon or doing field cultivations. Cows, however, are usually only employed for half a day. Being a dual purpose breed the Barrosa also fatten well.

All the old work oxen and breeding cows go to the butcher at the end of their productive life. The forequarters are not very tender but the back and loin is of good quality. Carcases are about 55% of the liveweight as the bone percentage is low. Young males achieve even better figures.

Galician Blond

In Spain, this breed is found in the provinces of La Coruna, Ponteverda, Lugo and Orense and was the only breed in these parts until the end of the 19th century. It is identical with the Galega, or Minho, breed found in the north-west corner of Portugal. Although treated as separate breeds in their respective countries, they are only kept apart by the Minhota river which borders Spain and Portugal.

The Galician Blond is a milk and meat-producing breed that is also used for work. In Portugal it is decreasing in numbers, as farmers prefer the better milk producing breeds. The Spanish animals have been crossed with Shorthorn, Brown Swiss and Simmental in order to improve their milk production. This means that only a few purebred animals still remain.

This breed has a thick, rather tight skin that is lightly pigmented. The coarse medium length hair is red but the depth of colour varies between animals in different areas. The male head has a convex forehead, whilst the females are flat. Muzzles are pink and large and the horns are yellowish with reddish or greenish tips. They grow out from the head before curving forward at the ends. The chest is deep and the ribs well sprung. Legs are strong and the light coloured hooves are hard. Although the forequarters are somewhat heavier than the hindquarters, these animals do produce good carcases when fed properly.

Normal liveweights of mature males are around 600 kg (1,323 lb) and they

Red Steppe (Russia)

Ala Tau (Russia)

Mongolian (Mongolia)

Miranda (Portugal)

Barrosa (Portugal)

Galician Blond (Spain)

Andalusian (Spain)

Fighting Bull (Spain)

average 139 cm (55 in) at the withers. Mature females weigh 410 kg (904 lb) on average and are 130 cm (51 in) at the withers. Where the cattle are grazing large areas of poor pasture they do not reach these weights, and even bone growth suffers.

A Herd Book was started in 1933 by the General Directorate for Animal Production, and is now maintained by the Spanish Government.

Andalusian

This breed has been developed from the local cattle that have for centuries inhabited the Spanish provinces of Córdoba, Seville and Badajoz. No outside blood was introduced until recent times, but now the breed is in danger of extinction due to the large amount of crossbreeding with imported stock.

The breed is known in Spain as the *Retinta Andaluza de la Cuenca del Guadalquivir*. It is found in the valleys of the Guadalquivir and Guadiana rivers, and in the country between. There are large areas of pastureland in this region and the grazing is good

from February through to June. After this comes a dry spell, which drastically reduces the herbage for two or three months before the autumn rains encourage more growth. During the drought, and in the winter months, the cattle lose weight which they put on again as the pastures recover.

Andalusian cattle are a dark mahogany red with soft, sleek hair that grows longer and rougher in the winter. The loose, medium thick skin is darkly pigmented. The long horns are the main physical feature of this breed. They are thick at the base and go out sideways from the head before turning forwards and then upwards. The tips are dark but the rest of the horn is a greenish-yellow colour.

These animals are dual purpose for work and meat. They have well developed muscles, short, strong legs and hard hooves. At two years old they are put to work, usually for about three years as they tend to get vicious with age. As youngsters they are willing workers and can be used for half a day at a time on about 150 days in the year. When mature the males average 650 kg (1,433 lb) and stand 141 cm (56 in) at the withers. Females are about 380 kg (838 lb) when mature and are 136 cm (54 in) at the withers.

Although the beef carcases produced from Andalusian cattle are not of the best quality, they can be fattened on grass alone. This gives them an advantage over some of the better European beef breeds.

Fighting Bull

Bred on the Iberian Peninsula, this breed is known as the *Brava* in Portugal and the *Toro de lidia* in Spain. Developed

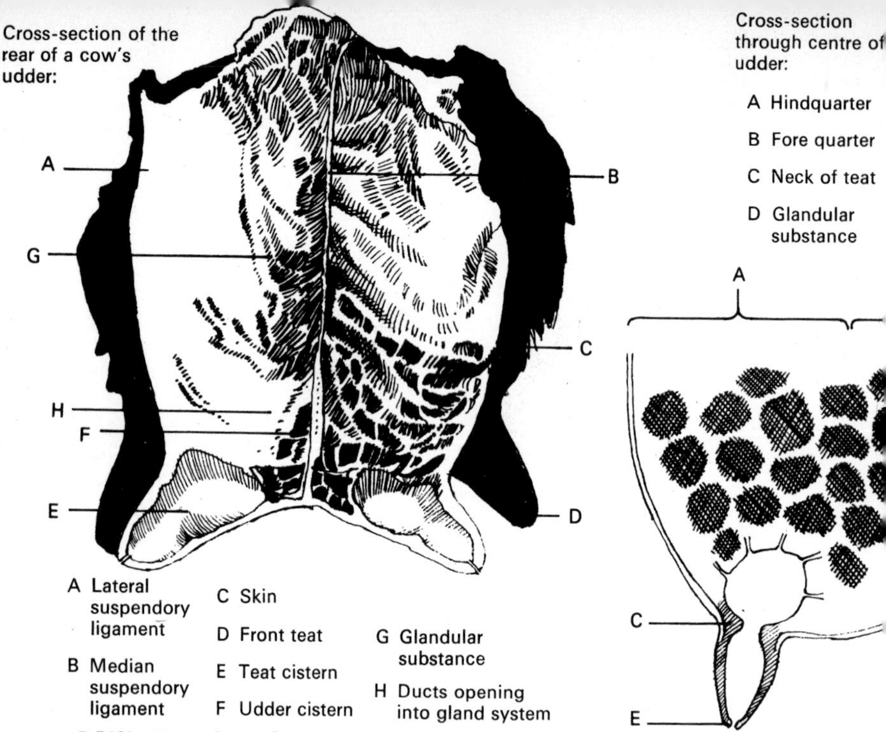

Milk Production

Mammals, when they give birth to their young, produce milk in their mammary glands. Although this milk is intended for their own young, the situation has been exploited by the dairy farmer. A cow's milk is diverted from her offspring to the family of man.

The Cow's Udder

The udder that carries the milk is situated between the hind legs and suspended by strong ligaments from the pelvis and muscular wall of the belly. It is composed of four separate quarters, or glands. Each mammary gland is made up of tiny cells which manufacture milk. These cells are grouped into hollow spheres called alveoli. Each alveolus has its own blood supply and draws chemicals from the blood in order to convert them into the quite different compounds present in milk.

Milk is secreted into the hollow centre of the alveolus and drains out through a tiny duct. This connects with other ducts before flowing into the udder cistern from which it can be drawn off through the teat. Within each teat is a small cavity called the teat cistern and the only barrier preventing the flow of milk from the udder is the teat sphincter muscle at the tip of the teat.

Milk Flow

When the milking of cows first began, it was realised that some form of stimulus was needed before the animal would let down its milk. If the cow's own calf was kept near her then

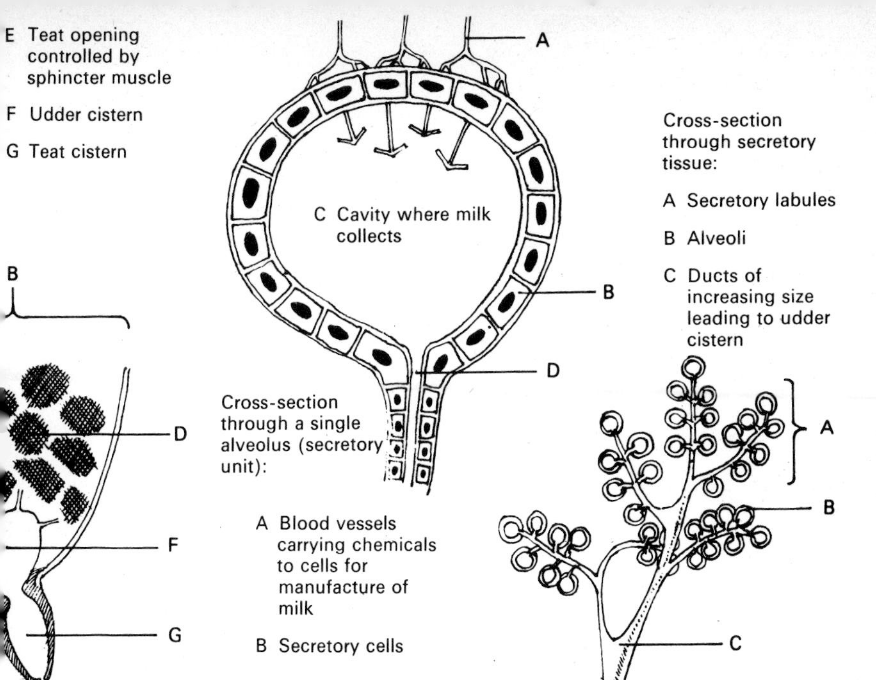

she would respond by allowing her milk to flow, even if the youngster was not suckling. A problem arose if the calf died, but this was overcome in a variety of ways. One method was to present another calf, smeared in urine from the dead animal, so that the cow would recognise the smell and adopt the calf. A second trick was to dress the substitute calf, or even a small boy, in the skin of the dead calf. The cow smelled the pelt, licked it, and soon submitted quickly to the milking routine. Although similar ruses are still used when getting cows to adopt calves, modern milkers no longer need to go to such lengths to obtain milk. Dairy cows today respond to other stimuli. The milk flow of some cows will be triggered off at the moment they enter the milking parlour. Others require to be fed and have their udders massaged by a washing technique.

When a cow is about to be milked, or suckle her calf, the hormone oxytocin is released from the posterior pituitary gland and causes tiny muscular cells surrounding the alveoli to contract. This squeezes the milk into the udder cistern. Emotional stress, which causes adrenalin to be secreted into the blood can quickly reverse the situation. The action of oxytocin is prevented by the adrenalin and the milk is not let down. This is a well known situation when young heifers come into milk for the first time and are not used to the milking routine. An experienced herdsman tries to minimise the stress factor and avoid upsetting his cows.

from ancient stock it is assumed that these cattle are directly descended from the wild aurochs.

In Spain there are more than two hundred ranches devoted to producing fighting bulls. Some of the largest are many thousands of acres of uncultivated land. The breeding cows roam free and are accompanied by the stud bulls from April to June. The calves are born in the open from December to February and suckle their mothers until weaning at eight months of age.

At each stage the animals are tested for bravery. The dams of the fighting bulls are expected to charge when annoyed and the stud bulls also have to show courage. Any beast that does not exhibit bravery, an aggressive character and fleetness of foot is not allowed to breed. When two years old, the young bulls themselves are tested and any that do not react aggressively are reared for beef. Those chosen for fighting are moved to better pastures and then sent to the bull-rings from three years onwards. They are lean and muscular and only weigh 700 kg (1,543 lb) on average. Fighting bulls are very athletic and can outrun a horse for 23 m (25 yds) and outmanoeuvre a polo pony. Their strength is incredible and they have been known to lift a horse and rider and toss them over their back.

The majority of the bulls of this breed are black but as colour is not a selection factor there are numerous animals of other colours. The relatively short head has strong, medium length horns that curve forward and end in a black tip. The short thick neck leads into the wide shoulders. Fighting bulls have very well developed forequarters and long bodies that are high at the withers. Legs are fine boned but strong, and end in small black hooves. Females of this breed are not as heavily built and have smaller heads with shorter and thinner horns. They have no visible udder and only produce sufficient milk for their calves.

About 3,500 bulls a year are sent to fight the matadors and provide a spectacle for the Spanish crowds and the tourists. In Spain the fight always results in the death of the bull, but in Portugal, where the bull's horns are either padded or tipped with a brass ball, he is not killed in the ring.

Bull fighting is a very ancient sport and when the Spaniards colonised parts of the New World they took with them this form of entertainment. The first recorded bull fight in Mexico took place as early as 1529. There are also bullrings in Bolivia, Colombia, Peru and Venezuela.

Africander

It is thought that the ancestors of the Africander were brought from Asia to Africa by the Semitic tribes about 1500 B.C. These *Bos indicus* cattle moved from Ethiopia down to the Great Lakes area. It was here that they came under the influence of the Hottentot people who migrated west and south to the Cape of Good Hope. The Portuguese first recorded seeing these Hottentot cattle in the 15th century and large herds were acquired by the first Dutch settlers in 1652. The subsequent development of these animals into the Africander breed was carried out by the European colonists.

Early records show that the Africander were good draught animals, having stamina, speed, hardiness and docility.

They could survive under dry, hot conditions on poor vegetation and were maintained in some remote districts to provide trek oxen. At one time they were the only means of transport throughout the greater part of southern Africa.

Rinderpest outbreaks, together with the Anglo-Boer war from 1899 to 1902, greatly reduced the numbers of Africander cattle. However, numbers then increased until alternative methods of transport and importations of European cattle threatened their existence. A Breed Society was formed in 1912 and breeders turned their attention to beef production. This proved successful and by 1951 about 30% of the cattle owned by Europeans in the Union of South Africa were of Africander type.

Mature males average about 907 kg (2,000 lb) and stand 142 cm (56 in) at the withers. Cows average 544 kg (1,200 lb) and 134 cm (53 in). The ratio of bone and fat to meat is comparatively low and the Africander provides a carcase that is about 60% of the liveweight.

Hottentot cattle were lean and long-legged, with considerable variation in type and colour. Since the Africander Cattle Breeders Society was formed selection has produced a uniformity of conformation and colour. The large head is coffin shaped when seen from the front with the widest part being between the eyes. The spreading horns are oval in cross section and turn downwards and backwards before curving upwards and forwards with a final backwards turn in mature animals.

The short, smooth hair is red and varies in shade between individuals. The amber, or brown, skin is thick, loose and supple. The long body has a firm, muscular hump in front of the withers which in the males is prominent and may have fat deposits under the skin and between the muscles. The dewlap is big and loose with folds that start under the neck and continue between the front legs.

In parts of South Africa, where the environment is suitable, European breeds of cattle can be kept. Where the conditions are harsh, then pure Africander or Africander crosses are reared for their resistance to the heat and disease.

Drakensberger

In 1659 the Dutch settlers in southern Africa found black native cattle owned by the Hottentots. During the early part of the 18th century these were improved by the use of Groningen bulls imported from Holland. The name 'Vaderlanders' was given to these animals by the Dutch and they were used as trek oxen in the great move northwards. Jakobus Johannes Uys and his son concentrated on improving the 'Vaderlander' type by breeding and selection. The farmers in their district referred to these cattle as 'Uysbeeste' after this family of trekkers. The Minister of Agriculture officially recognised this breed in 1947 and changed the name to Drakensberger.

These large, black animals have a loose, darkly pigmented skin and a sleek coat. They have short, white horns with dark tips that grow straight out from the head. The body is long, deep and broad with only the males having a small shoulder hump. Legs are sturdy and hooves strong. Cows have small but well-shaped udders and the breed is considered to be triple purpose for milk, meat and draught.

Drakensberger breeders are concentrating on producing beef animals that will grow fast and convert food efficiently into meat. At the same time they do not want to lose the hardiness that enables their breed to survive in extreme climatic conditions. In the Drakensberg area of Natal and Orange Free State the winter temperatures can drop below freezing point whilst the scorching summer heat rises to 43°C.

Mature males should weigh between 815 and 910 kg (1,800–2,000 lb) and cows have recorded weights up to 710 kg (1,560 lb).

Nguni

The now extinct Hamitic Longhorn of north-eastern Africa is thought to have been an ancestor of the Nguni. Way back in antiquity, crossbreeding between the Hamitic Longhorn and the Indian zebu produced a type of cattle that has been classified as 'Sanga'. These animals were taken by the Bantu tribes on their migrations down the east side of Africa and reached the south in about A.D. 500. Originally called 'Zulu' or 'Swazi' cattle, according to which tribe owned them, they are now officially known as Nguni after the name of the tribal group. Similar cattle in the south of Mozambique are called 'Landim'.

The Nguni cattle play a big part in the economic and social life of the natives. They are primarily kept for milk but those that die naturally are eaten. Hides are used for making many things including shields and loin aprons. Oxen are used for draught purposes.

Wealth is measured by the number of cattle a tribesman owns. Zulu and Swazi pay for their wives with cattle, according to the custom of *Lobolo*. Normally, thirteen head of cattle is paid for a girl who is not of royal descent. One is given to her mother 'to wipe away her tears', two are slaughtered during the wedding festivities and the remaining ten become the property of the bride's guardian.

Tribal custom has tended to prevent the development of the breed as numbers of the cattle have been the objective rather than quality of the individual animal. This has led to overstocking of the available pastures and lean and hungry animals. However the Kwa Zulu Government is now breeding an improved strain of Nguni cattle for resale to Africans in order to increase the performance of their stock. Initial

selection was on milk production alone but now fertility and weaning weights of calves are also being taken into account.

The short, fine, glossy coat of the Nguni covers a pigmented skin. The hairs come in a variety of colours which are either whole or mixed. White, black, brown, red, fawn and yellow animals occur and there are seven main pattern markings all with Zulu names. These patterns are used to great effect in the design of Zulu shields.

Nguni are small, dairylike animals with mature bulls averaging about 545 kg (1,200 lb) and cows 363 kg (800 lb) but weight varies greatly with environment. In some overstocked areas cows may weigh no more than 227 kg (500 lb) and stand only 107 cm (42 in) at the withers.

In cows, the horns are characteristically lyre-shaped and point upward and slightly forward. Bulls horns are shorter, stouter and crescent shaped. The tips are always dark and usually the rest of the horn is dark to match the hooves and muzzles. The muscular hump is well developed in bulls and hardly noticeable in cows. Udders are small to medium in size and usually have small, pigmented teats. Good cows can yield 9 kg (20 lb) of milk each day.

Bapedi

The Bapedi tribe of Nguni Bantu live in the Sekhukhuneland district of Eastern Transvaal. Their cattle are regarded as a type of Nguni but are called Pedi, or Bapedi, after the tribe.

In the 1930's, the finest Bapedi cattle were in the herd of Chief Ntoampe Mampuru. When the South African Bantu Trust wanted to start a breeding herd, they were allowed to select individual animals from those owned by Chief Mampuru's herdsman. In return, he was granted free grazing rights on nearby farms and an equivalent number of cattle to those taken by the Trust.

The animals originally selected for the Government herd were chosen on colour. The predominant colour is black with white on the belly and lower neck border. There is also a blue roan formed by the invasion of white hair into the black area and some animals are nearly all white. All black cattle were eliminated as being undesirable and possibly having some Africander blood. Also, animals with more than a spot or two of red colour in their coats were rejected—although, traditionally, small red patches in the coat were held in high regard by the Bantu.

Bapedi cows have straight heads whereas the males have concave heads with shorter thicker horns. Naturally polled animals do occur and also some have scurs. The hump varies considerably in size and shape but the dewlap is always well developed. Legs are fairly long and rumps are sloping. Cows have small udders and teats.

Adult bulls can weigh up to 680 kg (1,500 lb) and cows between 360–520 kg (794–1,147 lb). This is a good size for native cattle and there is quite a demand for bulls from Bantu farmers.

Throughout the improvement programme carried out by the South African Bantu Trust attention has been paid to regularity of calving, milk production, conformation and resistance to tick-borne diseases. Now growth rate is being measured, as a performance testing scheme was started in 1972.

Boran

The nomadic Borana people of the Lake Rudolph region have given their name to these African cattle. The three countries of Ethiopia, Somalia and Kenya all have Boran cattle, but the most highly developed are those in Kenya. The Kenya Boran are larger and of better conformation than the other varieties.

Somali traders originally brought the Boran cattle to Kenya and sold them to the Samburu, Masai and Meru tribes, as well as to the European settlers. Most of the improvement to the breed has taken place since 1920 and although this has mainly been by selection and breeding some crossing with European breeds has taken place.

The improved Boran is a beef breed and carcases compare favourably with those of European beef breeds and with the best Africander herds. One disadvantage is that the Boran is slower maturing and so it takes considerably longer to produce finished beef.

Boran are normally white with a thin pigmented skin. Grey, fawn and red animals occur, but black is rare and not considered a true Boran colour. The head is long and tends to be coffin-shaped with a broad muzzle. Horns are usually short, round in cross section and upright. However polled animals are common and some herds have been selected for this factor. The hump is well defined and larger in the male, sometimes leaning over to the rear. Behind the hump the Boran has an unusually level topline for a *Bos indicus* or 'zebu' type. Also the rumps are long, wide and muscular with the upper thigh thick and rounded. The long tail is set low and the dewlap is well developed.

There is considerable variation in the size of Boran cattle, but mature bulls normally weigh between 550 and 675 kg (1,210–1,490 lb) and mature females from 350 to 450 kg (770–990 lb).

Ankole

Ankole is the name given to a 'Sanga' type of cattle found in Uganda and neighbouring countries. The biggest concentration of Ankole inhabit the area between Lake Albert in the north and the northern tip of Lake Tanganyika in the south. There are a number of varieties of Ankole cattle and they can be conveniently classified according to the tribes who own them.

The largest, long-legged variety, is called the Bahima after the Hamitic pastoral tribe of the same name. The favourite, and most common colour of these animals is deep red, but red and white, dun, black, black and white and greyish white also occur. Horns are very long and large and normally curve outwards and upwards before turning inwards at the tips.

Somewhat smaller than the Bahima are the Watusi cattle, with their short soft hair and darkly pigmented skin. In colour they are brown, red, black or combinations of these colours with white. There are a variety of horn shapes and sizes and one strain, the Inyambo, with exceptionally long horns, is considered sacred. Measurements of 2·3 m between the points have been recorded. Polled animals and those with loose horns growing downwards are also seen.

Bashi cattle are found to the west and south-west of Lake Kivu and belong to the Bashi tribe of Bantu. These people are more farmers than herders of cattle

Africander (South Africa)

Drakensberger (South Africa)

Nguni (South Africa)

Bapedi (South Africa)

Boran (Kenya)

Ankole (Uganda)

Kuri (Chad)

White Fulani (Nigeria)

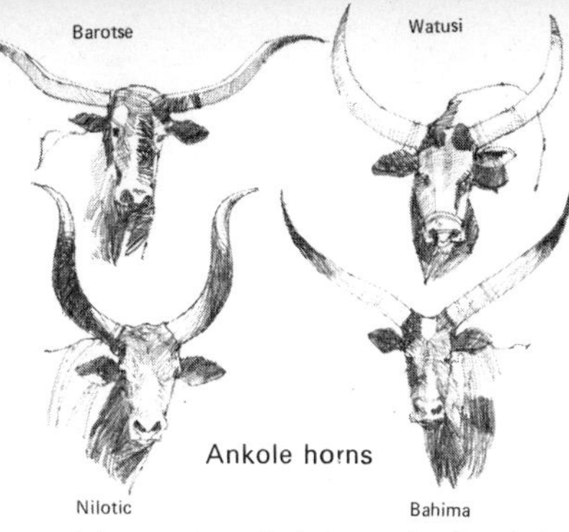

Ankole horns

and have selected their cattle for their small size and relatively short horns. Adult males are only 290 kg (640 lb) on average and females about 240 kg (530 lb). This is a lot less than the Bahima variety where the males are about 500 kg (1,100 lb) and the females around 340 kg (750 lb). The colour of the Bashi varies from red to fawn or black and often some white is mixed in with one of these colours.

Similar to the Bashi cattle, but slightly larger are the Kigezi from the south western corner of Uganda. They have shorter and more upright horns than the Bahima and tend to be paler in colour than other Ankole types.

All the tribes that breed Ankole cattle treat numbers of animals as an indication of wealth. More recently they are beginning to see the value of their cattle as producers of milk and meat. Where this is happening, the Ankole cattle are declining in numbers in favour of the 'zebu' types that are more resistant to disease and better producers.

Ankole cattle are still important in tribal ceremonies and are ritually slaughtered. Also some tribes bleed them as blood is a part of their diet. Very little use is made of them as draught animals but in some areas they are put to work at 30 months. They can toil for up to five hours a day ploughing the hard dry soil.

Kuri

These are the largest cattle in West Africa and are unique to that area. They are also called Lake Chad cattle or Buduma as they are kept by the closely related Kuri and Buduma tribes that live on the islands and shores of Lake Chad. Herds usually consisting of about 35 cows and a bull can be found grazing on the lakeside grasses. The animals swim from island to island in search of food and spend several hours in the water each day. They do not appear to thrive as well when moved away from the lake.

Kuri are tall cattle with long heads and gigantic horns that are porous and therefore not heavy. Normally, the horns are long and circular in cross section, but shorter, flatter horns do occur that give the appearance of enormous ears. Polled animals are present in small numbers in the population and some have horns that are not fixed firmly to the head but appear as long, loose scurs. The short neck leads into a long body with no hump. At the bottom of their long legs are large, open hooves. Coat colour is generally white but some animals are grey over the shoulders and others have red patches on the body.

Mature males weigh about 650 kg (1,430 lb) and stand 146 cm (57 in) at the withers. Cows when mature weigh about 400 kg (880 lb) and are 136 cm (54 in) at the withers.

The breed is used mainly for milk and meat production although sometimes as a pack animal. For draught purposes it is slow and tires easily.

Bulls are used when three years old and have a breeding life of about nine years. Cows calve for the first time at between 42 and 48 months of age and then produce calves at 15 to 18 month intervals.

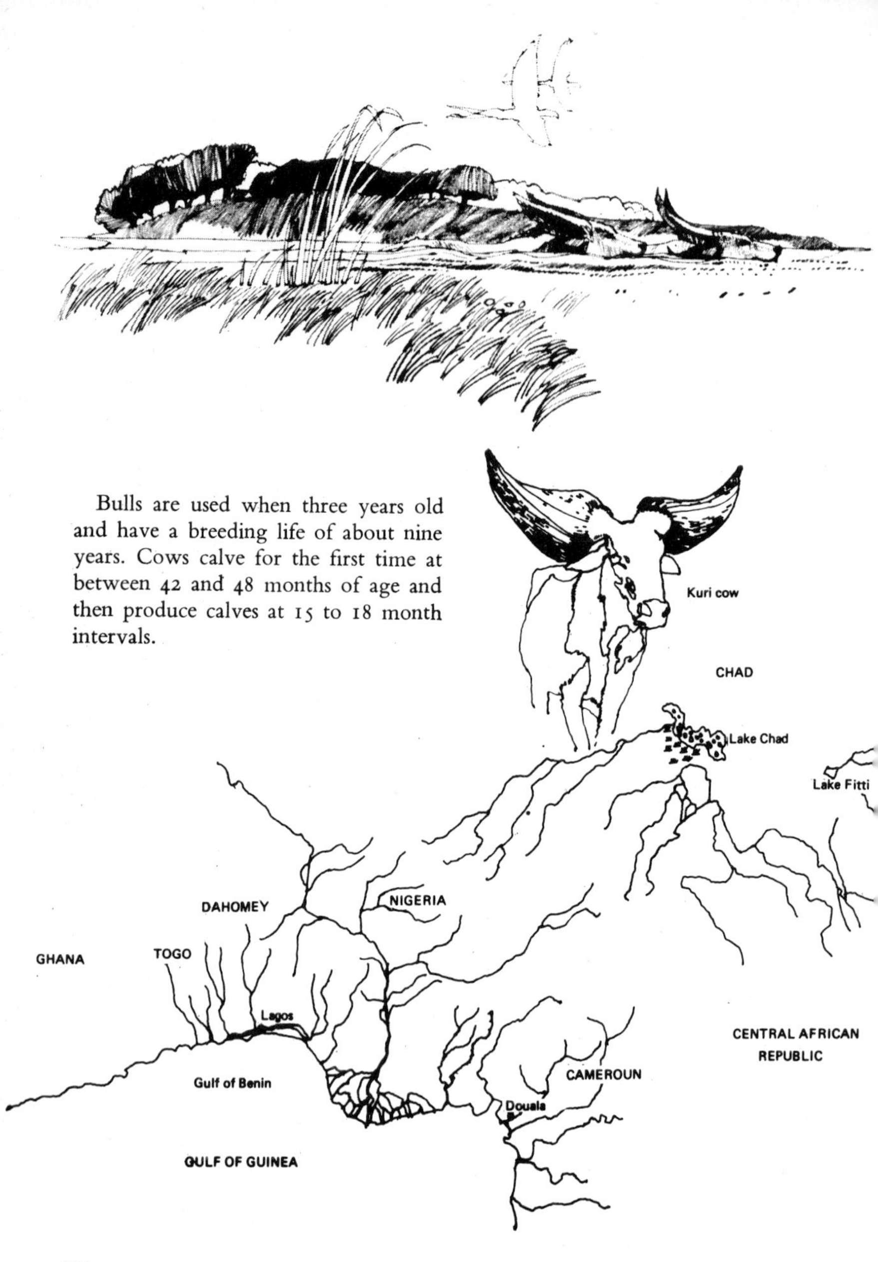

White Fulani

These lyre-horned 'zebu' type cattle take their name from the nomadic Fulani tribe. There are four varieties of these cattle, the largest being the White Fulani of northern Nigeria. The ancestors of these cattle are thought to have come from Asia and interbred with already established types such as the Hamitic Longhorn.

Although the usual colour of these cattle is white, there are also animals with black, and red, markings. The skin has a black pigment and the ears, eyes, muzzle, tail switch, hooves and tips of the horns are also black. The long face has a wide forehead and broad muzzle. Soft hair covers the loose skin, and the hump and dewlap are both well developed. Cows have good sized udders with medium sized teats.

Normally, the men of the tribe own the cattle but the women claim the milk and butter, which they then use as part of their daily food supply. Any surplus is exchanged for grain grown by those tribesmen that have settled down to farm the land. These farmers also buy beasts for fattening, as well as work oxen. The White Fulani are, therefore, a triple-purpose breed—although they are only slow draught animals. They walk at about 3 km/hr (2 miles/hr), and can keep this up for a six hour day, carrying out various field and haulage operations.

This breed responds well to fattening and castrated males are ready for the butcher at about five years of age. The best males are retained for breeding purposes and the tribesman are very reluctant to part with any cows until they are no longer fertile.

Mature males weigh 545 kg (1,200 lb) on average and stand 134 cm (53 in) at the withers. Females when mature average 340 kg (750 lb) and stand 127 cm (50 in) at the withers.

N'Dama

N'Dama when translated means 'small cattle' and it is thought that this breed descended from the animals that migrated south with the Berber tribes from Morocco. The centre of development of the N'Dama type was the Fouta Djallon plateau in Guinea. From here the breed has spread out to various parts of West Africa and has been introduced into Nigeria and Ghana.

This breed greatly increased in numbers after the serious rinderpest outbreaks of 1890–1891 and 1918, when other breeds were drastically reduced by the epidemics. Although susceptible to rinderpest and tuberculosis, the N'Dama's have a certain resistance to trypanosomiasis, which is carried by the tsetse fly. They can, therefore, be kept in tsetse bush country where other cattle types, such as 'zebus', do not thrive as well.

The N'Dama are humpless cattle, small and sturdy, with only a small dewlap. On top of the short, broad head they carry long lyre-shaped horns

Herringbone milking parlour – 16 stalls, 8 units

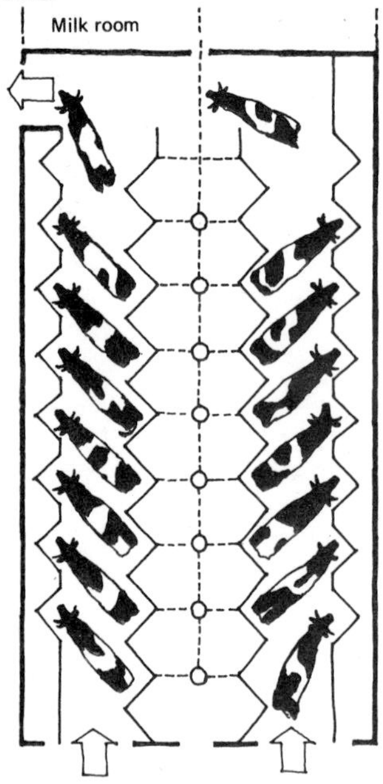

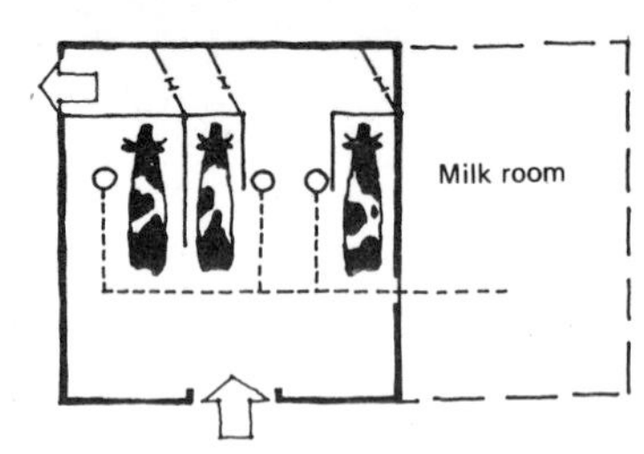

Abreast milking parlour – 3 stalls, 3 units

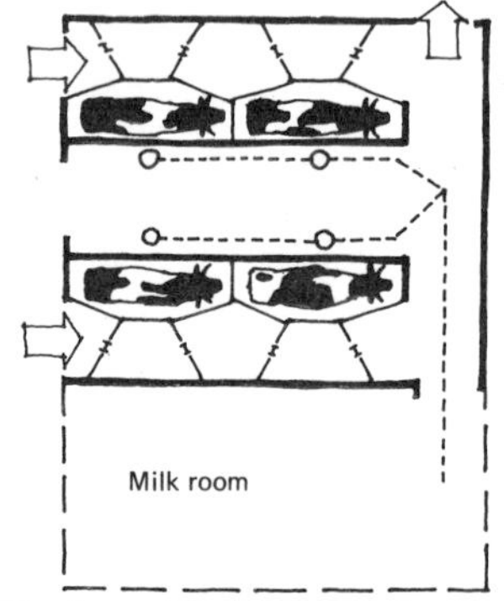

Tandem milking parlour – 4 stalls, 4 units

Milking Techniques

During the last 125 years there has been great technical developments in the dairying industry. Milking machines using vacuum suction techniques have replaced hand milking in the developed countries and a modern milking parlour is a highly efficient production line. The first patent for a milking machine was taken out in America in 1849. It proved popular as it had the double advantage of being faster and more hygienic than hand milking.

Hand milkers used a bucket or other container into which they directed the milk. The contents were later transferred into a churn. Originally, the churns were made of wood, but now are all metal. Early milking machines also deposited the milk in buckets, but before long systems were devised which piped the milk direct from the cow to the churn. This cut down the handling problems, but still left the milker to move the heavy churns to where a lorry could come and collect them. Now a system has been designed where the milk flows through a pipeline from the milking machine to a refrigerated tank. It is stored in bulk until sucked out through a hose into a milk tanker lorry.

Developments in machinery used for milking cows paralleled advances in the design of milking parlours. Originally, cows were milked where they happened to be. Usually, this was in a field in the summer and in a shed in

Rotary milking parlour – 14 stalls, 14 units

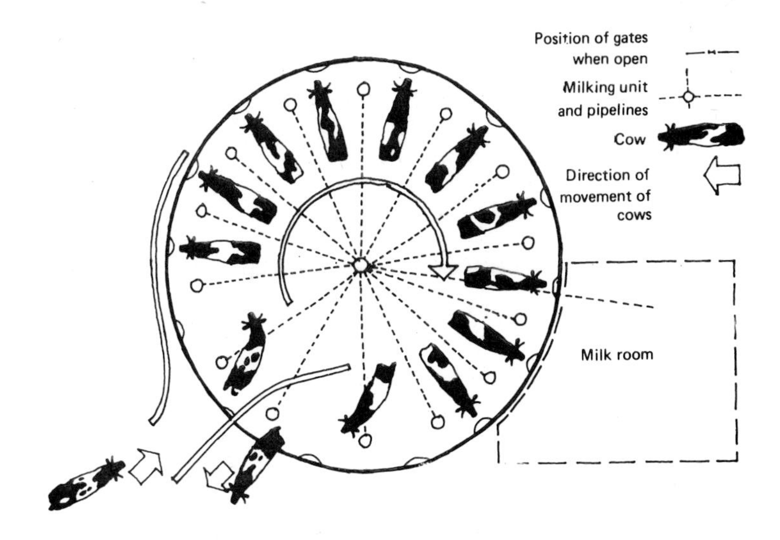

the winter, but some of the cows belonging to the London town dairies of the late 19th century were milked in the streets.

The first buildings for dairy cows had a space for each animal, where it could be tied up and milked. This meant that the milker had to move from cow to cow. Labour-saving milking parlours were built so that the cow moved into one of a limited number of spaces near the milker and then walked out again after being milked. These constructions normally conformed to one of three basic designs:

Abreast parlour Cows stand side by side with their backs to the milker.

Tandem parlour Cows stand in line one behind the other with their sides to the milker.

Herringbone parlour Cows are at a 45° angle to the milker.

Technology has recently produced a highly mechanised way to speed up the milking routine namely the *Rotary parlour.* As a cow steps on to a rotating platform she is identified by the milker and her number punched into a computer which activates a feed hopper and chute to give the correct amount of supplementary food. The cow has her udder washed and teat cups put on by the person milking. By the time she has been taken on one circuit of the parlour, she has been milked out and the teat cups have been automatically removed.

that grow out and slightly backwards, before turning upwards and inwards with backward pointing sharp tips. The soft, hairy coat is short and in the Fouta Djallon area is normally a fawn colour. In other parts of West Africa, the colour varies from white through yellow to brown and red. The skin pigmentation also varies from light to almost black.

There is a wide range of wither heights and mature weights depending on location. Bulls are not normally less than 95 cm (37 in) at the withers and not more than 118 cm (46 in), whilst cows are slightly shorter. Bulls can weigh from 300–420 kg (645–925 lb) and cows from 230–350 kg (505–770 lb).

Milk yields from the N'Damas is poor and although they are used for work, they cannot keep up the effort for long periods. Their best use is as beef cattle and carcases are produced that give 45–55% of their weight as usable meat.

Egyptian

There is no distinct type of Egyptian cattle, but instead a number of similar varieties in different parts of Egypt. These strains are sufficiently alike to be considered under the same heading.

In southern, or Upper Egypt, along the Nile valley, are the Saidi type; in the Western Desert are the Maryuti that belong to the nomadic tribes. Lower Egypt has two local types, the Damietta by the coast and the Baladi further inland.

Shortage of water is a great problem in Egypt, for apart from the few centimetres of rain that fall each year along the northern coast, the rest of the country has a negligible amount. Irrigation is therefore essential in order to grow cotton, maize, rice and other crops. At one time, nearly all the power for the irrigation and cultivation was supplied by native cattle and buffalo. Now the draught animals are being replaced by machinery and imported European cattle provide the milk and beef, which is supplemented by buffalo milk and camel meat.

The coat colour of Egyptian cattle varies from fawn to red. The narrow head is medium in length and the short horns grow straight out from the flat poll before curving forwards. The long, lean body is not very deep and the ribs are flat. Toplines are uneven being low in the loin and high at the tail head. Only the females of the Saidi type show any signs of a hump, although all bulls have a prominent crest.

Kankrej

Kankrej cattle provide the heavy, draught animals of India. They are considered to be the largest and most powerful working oxen in the subcontinent of India.

The breed has been kept pure for many generations by the semi-nomadic peoples that live to the north of Bombay. The home territory of the Kankrej stretches from the south-west corner of the Tharparkar district of Sind, to Dholka in the Ahmedabad district in the south, and from Dessa in the east to the extreme end of the Radhanpur State in the west.

Many animals have been exported to other parts of India for draught purposes and also to other countries of the world. Around 1900, Brazil bought approximately 400 Kankrej cattle from the Indian villagers. Some were re-exported to the U.S.A., where they are known as the Guzerat and have helped form the American Brahman breed.

The head on Kankrej animals is carried high and bears long lyre-shaped horns. These can be induced to grow thicker by removing the coronary ring and the outside layer of horn. In the villages this is done by a specialist who bites the horn buds of the calves, whereas on the farms a knife is used when the animals are about two years old.

Other characteristics of Kankrej cattle are their drooping ears, well developed hump, moderately developed dewlap, thick skin and black tail switch. The rest of the body is grey. Some calves are born with reddish coats but they turn grey by the time they are five months.

Over the years selection has been for alert animals that can walk well and quickly. Now the oxen are highly prized as draught animals and a well-broken, four-year-old pair will fetch a high price. Adult bulls in good condition can weigh 726 kg (1,600 lb) and stand 158 cm (62 in) behind the hump. Adult females would weigh about 454 kg (1,000 lb) and stand 150 cm (59 in).

Gir

This Indian breed was developed in the Gir forest of southern Kathiawar. The area covers some fifteen hundred square kilometres on the rectangular peninsula between the Gulf of Cutch and the Gulf of Cambay on the west coast of India. The rainfall varies from 51 cm (20 in) to 102 cm (40 in) and therefore the cattle are not short of water. There is also unlimited grass in the shaded forest.

Gir cattle have characteristic domed foreheads and horns that curve back and then upwards. They are usually about 5 cm (2 in) in diameter and 30 cm (12 in) long, although some animals have loose scurs. Ears are long and pendulous, opening to the front and looking like a curled up leaf, with a notch near the tip.

Some animals are entirely red but most are mottled and vary from yellowish-red to almost black. A peculiarity of the breed is that most animals have a well-defined patch on them that differs markedly in colour from the other areas.

The cattle are deep bodied and with fairly straight backs behind the hump. The dewlap is only moderately developed. Mature males weigh about 658 kg (1,450 lb) and stand 152 cm (60 in) at the withers. Females when mature weigh about 386 kg (850 lb) and stand 142 cm (56 in) at the withers.

The Gir is a milking breed and is thought of as a dairy type 'zebu'. Castrated males are used for draught purposes, but although they are big and powerful, they tend to be slow and lethargic.

New breeds of cattle have been developed in both North and South America using Gir blood. Between 1875 and 1921, Brazil imported Gir and Kankrej animals from India and formed the Indo-Brazilian breed. In the U.S.A., the Gir (along with three other *Bos indicus* breeds) were combined into the Brahman breed.

Khillari

The Khillari are Indian draught 'zebu' that are included in a group of cattle called the 'Mysore' type. It is thought that the breed is descended from the Amrit Mahal cattle that were specially bred for warfare, from about 1600, by the former rulers of Mysore State. Compact, active and with a fiery temperament the Amrit Mahal draught animals were famous for their powers of endurance. At the very beginning of the 19th century when Hyder Ali, the ruler of Mysore, was fighting the British, he managed to move his army 160 km (100 miles) in 2½ days, using Amrit Mahal cattle to pull the guns and equipment.

Khillari cattle are found mainly in the Sholapur and Satara districts of Maharashtra State, to the south-east of Bombay. They are not so compact or as active as the Amrit Mahal and there is a wide variation of type. Some distinct varieties have their own names, such as Atpadi Mahal, Mhaswad, Nakali and White Thillari. All are grey or white, with very prominent foreheads and long faces tapering towards the muzzle. Their long, pointed horns emerge near each other at the top of the head and curve backwards and upwards in a graceful sweep. These are the characteristics of the 'Mysore' type of cattle, as is the tight skin stretched over the body, which is unlike a number of other 'zebu' that have loose skin in folds. The Khillari cows are poor milkers and it is noticeable that the 'zebu' strains with tight skins are the low milk producers but fast draught animals. At the other extreme are the loose skinned 'zebu' that milk well but make slow draught animals.

Tharparkar

Tharparkar cattle are a dual purpose 'zebu' for draught and milk. They are found in the arid, semi-desert area of south-west Sind Province in Pakistan where the local name for them is *Thari*. Some have spread into adjoining areas of Jodhpur and Cutch in India. However, there are not many pure Tharparkars in the Cutch, as breeders there are concentrating more on the Kankrej cattle, because of the greater demand for draught oxen of that breed.

The area in which the Tharparkars are bred is very sparsely populated, with isolated villages built around wells. The northern boundary extends to the

N'Dama (Guinea)

Egyptian (Egypt)

Pedigree cattle being auctioned at

Breeder's Trophy
and rosettes

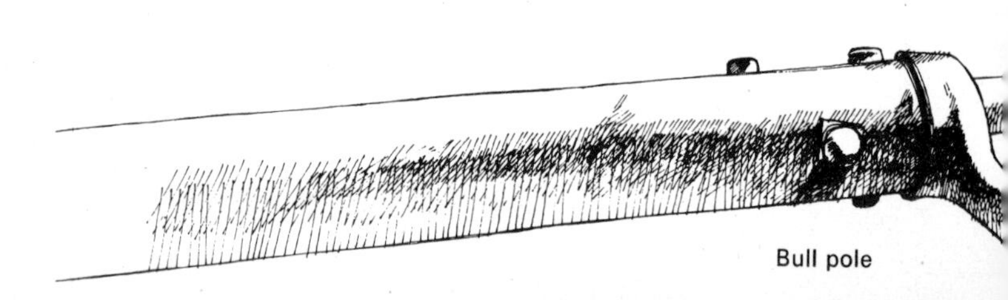

Bull pole

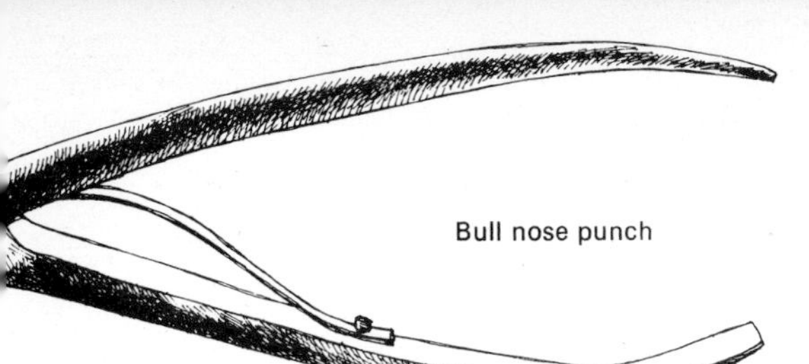

Bull nose punch

:ish Charolais Society's Show and Sale

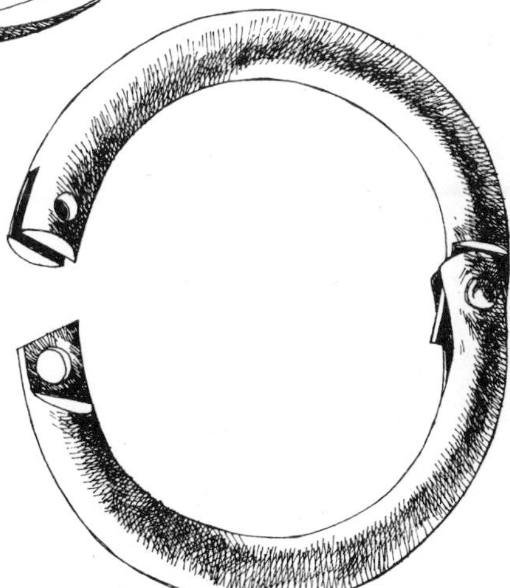

Ordinary nose ring

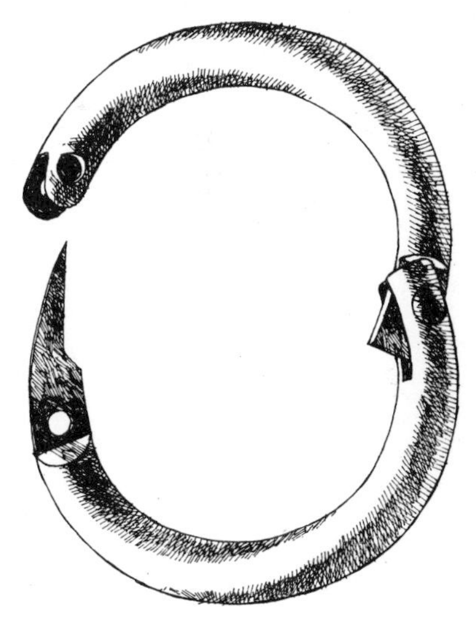

Self-piercing nose ring

Kankrej (India)

Gir (India)

Thar Desert and cattle have to be moved over long distances in search of food. Having lived for centuries in these near desert conditions, the breed is very hardy and heat resistant.

Tharparkar are medium size, compact cattle which if fed properly, develop into deep bodied animals. Adult cows stand between 135 cm (53 in) and 142 cm (56 in) when measured behind the hump and weigh about 408 kg (900 lb). Adult bulls are nearer 150 cm (59in) and reach a weight of 612 kg (1,350). Horns are medium small and turn up before curving in at the tips. Humps and dewlaps are reasonably well developed and the skin is fairly tight. Hooves are strong and black.

Threequarters of all Tharparkar cattle are white to blue-grey, with calves having a red tinge. Most of the remainder are red and this is attributed to the influence of the Red Sindhi breed on the western border of the Tharparkar area. In the east and south, there is evidence of Kankrej blood.

Male Tharparkars that are castrated and used for draught purposes are fast, willing workers. On metalled roads they can maintain a speed of almost 10 kilometres/hour (6 miles/hour) for a considerable distance. In the fields, they work for six to ten hours a day and carry out various operations such as ploughing, threshing and working wheels or sugar cane crushers.

The cows, although temperamental, are reasonable milk producers that can yield up to 4,536 kg (10,000 lb) of milk in a lactation, under correct management conditions. However, more normal lactations vary from 454 kg (1,000 lb) to 3,402 kg (7,500 lb), at approximately 4·3% fat. Yields are lower in the desert areas where the cattle live in an almost wild state. They remain in the open all year round and come back to the villages each morning for watering and to be milked. Young animals are kept in the villages from birth until six months, when they are weaned and grazed separately, so that they do not suckle their mothers.

Red Sindhi

The home of this breed is the near-desert country of Sind Province in Pakistan, in the districts of Karachi and Hyderabad and in the Las Bela area of Baluchistan.

Various countries have imported the breed as it is very hardy and can adapt to different climates. Herds have been started in Malaya, Burma, Sri Lanka, the Philippines, Japan and Brazil. Numbers of bulls were shipped from Karachi to Singapore, standing on matting placed on the iron decks. They were tied to the bulwarks for the whole journey, and although they had no shelter through the tropics, none died.

Red Sindhi are small and a deep red colour, which has led to the belief that they are descended from the red cattle of Afghanistan. Some of the animals have white patches on the face and dewlap, and occasionally blue-grey calves are born to Red Sindhi parents.

The relatively short horns of the Red Sindhi turn upwards. Bodies are deep, ribs well sprung and rumps slope. Cows have well developed udders and the breed is considered to be one of the better milking strains of 'zebu'. A good cow should give about 2,040 kg (4,500 lb) of milk in 305 days.

When fed reasonably well, adult males should weigh about 544 kg (1,200 lb) and adult females 340 kg (750 lb). Height just behind the hump is 127 cm (50 in) for adult males and 112 cm (44 in) for females.

Texas Longhorn

Longhorn cattle were introduced to the American continent by the Spaniards early in the 16th century. The small, thin animals thrived on the Mexican plains and in the area that is now the State of Texas.

The only cattle in America before the arrival of Hernando Cortes and his Conquistadores in Mexico were the wild Bison. Then, two years after the Spanish conquest of 1519, the first few long-horned animals were shipped from Spain. This was followed by other importations and many ranches were established. Cattle numbers increased and the size of the animals was improved.

Texas remained a part of Mexico until 1836, when the Anglo-American settlers defeated the Mexican army and declared Texas a republic. Large numbers of Mexican ranchers returned home to Mexico, but most left their cattle behind. The ranches and the cattle were taken over by the Texans and eventually there were too many animals for the home market. Some ranchers tried to drive their animals to the northern cities of America where the meat was needed. Many men and beasts died in the attempt, as the distances were so great and the journey so difficult.

All cattle trade with the North was halted when the Civil War began. Then, in 1866, when the war ended, the cattle drives began again, as the Texans were short of money and had thousands of cattle to sell. The building of a railway line from Kansas to Abilene shortened the distance the cattle had to travel. The trail from the south to Abilene was some 900 miles across Indian country and drought areas. This was the famous Chisholm Trail and ran almost parallel to the Western Trail, which ran into Dodge City.

Texas Longhorns are all colours and patterns, with red being the most common colour. Their characteristic long horns are usually some 1·5 m (4 ft) from point to point with some animals recorded at 2·4 m (7 ft).

This breed was crossed with, and then replaced by, Hereford and Shorthorn cattle after about 1880 and almost became extinct. However, a Breed Society was formed in 1964 and now a number of herds are still to be found in Texas, including the one on the 3S Ranch at D'Hanis owned by Alan Sparger and Son. Both colour pictures on pages 152 and 153 were taken by Alan Sparger III and show 'Hondo' the main herd sire, and one of the best cows.

Brahman

The humped cattle of India are of the species *Bos indicus* but are commonly called 'zebu'. In America, they are referred to as Brahman. Indian cattle were first imported into the U.S.A. in 1849 by Dr. James Bolton Davis of Fairfield County, South Carolina. It is thought that Dr. Davis became aware of the potential of 'zebu' cattle when acting as agricultural advisor to the Sultan of Turkey.

In 1854, two Indian bulls were presented to Mr. Richard Barrow of St. Francisville, Louisiana by the British Government in recognition of his services in teaching cotton and sugar cane culture to a British representative who was to establish these crops in India. These bulls were used for draught purposes but also sired some cross-bred progeny known as 'Barrow Grade' cattle. The fame of these two bulls and their offspring sparked off a series of importations. Quite a large number of cattle came via Brazil, as Indian animals had been imported there from 1870 onwards.

There are more than thirty breeds or varieties of Indian cattle but the majority of the animals brought to the U.S.A. were of four strains. The Kankrej were the largest and quietest, varying in colour from almost white to dark iron-grey or black. Darker areas usually appear on the forequarters and hips. In the U.S.A. this strain has become known as the Guzerat.

The Nellore variety is also large for an Indian breed and the cattle are steel-grey to white in colour. Their horns are inclined backwards and upwards and their smaller pointed ears do not droop as much as other 'zebu'.

The third grey-white type, called Krishna Valley, is similar to the Nellore, whilst the Gir is a different colour, being yellowish-red to almost black. Many Gir have white markings and can have a roan appearance. They have large drooping ears and more pronounced dewlaps than the other three varieties.

Between 1910 and 1920, many of the cattle in south-west Texas and the coastal part of the Gulf of Mexico showed considerable 'zebu' influence. Many breeders tried to keep the imported cattle pure but the majority of the bulls used in this area were crossbred.

Mr J. W. Sartwelle of Houston, Texas, who proposed the name Brahman, was the first secretary of the American Brahman Breeders Association, organised in 1924. A Herd Book was established and the first cattle were registered.

The short, thick, glossy coat of the Brahman covers a black pigmented, loose skin. The preferred colour is steel-grey but some breeders are selecting for a solid red colour and this is becoming more popular. The long head carries short horns that turn up. Bodies are long and deep and considerably lower set than the Indian cattle originally imported. The hump is pronounced in both males and females. Brahman cattle continue to grow until five or six years of age, when bulls, on average, weigh 816 kg (1,800 lb) and cows 544 kg (1,200 lb).

Heat-tolerance and disease-resistance are two very important factors in the success of the Brahman in the hot south of the U.S.A. A third factor is the hybrid vigour obtained by crossing Brahman cattle with the unrelated European

Cattle Rail Transport

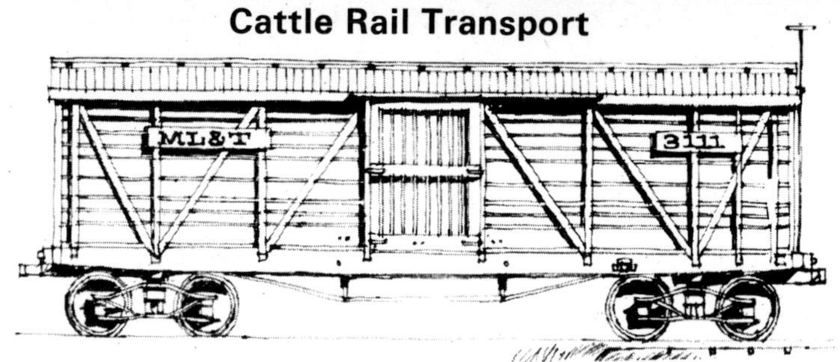

American stockcar of Morgan's Louisiana and Texas Railroad, 1887

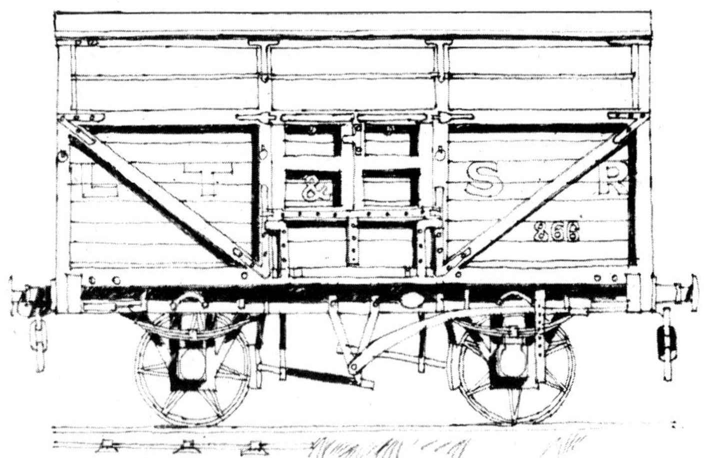

British cattletruck of 1903

American stockcar of the St. Louis, Iron Mountain and Southern Railroad, 1922

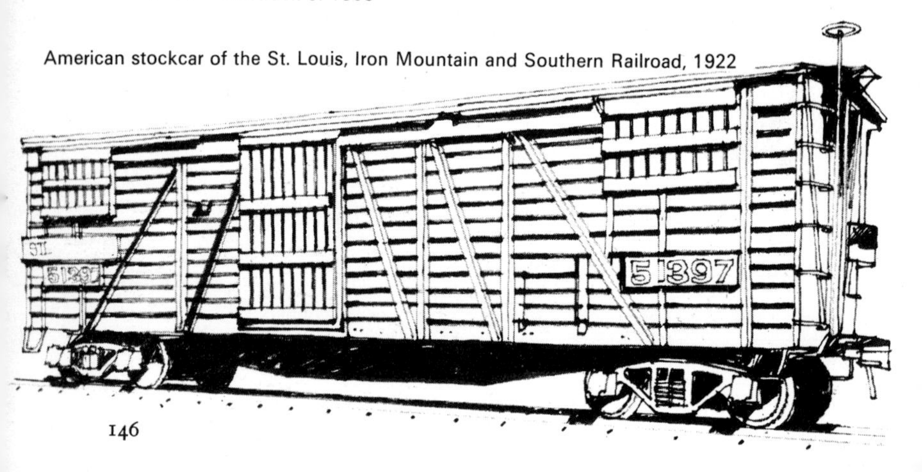

Milk Rail Transport

:ish milktank wagon of the South Western Railway, pre-1939

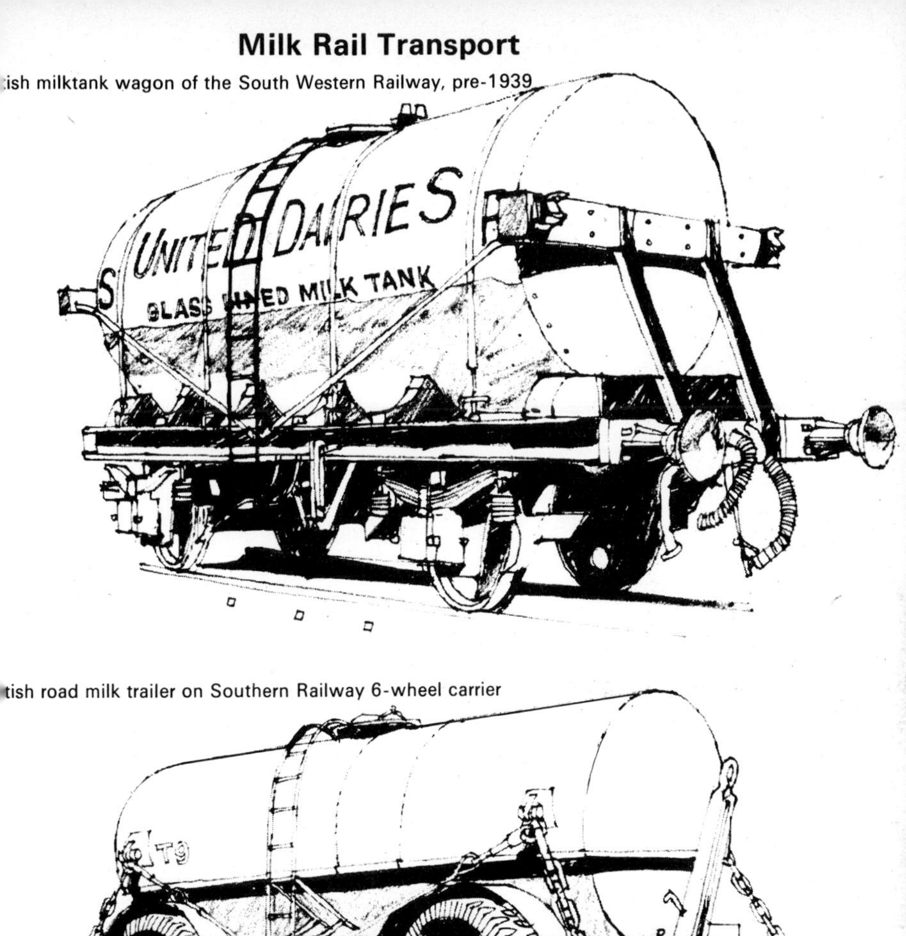

tish road milk trailer on Southern Railway 6-wheel carrier

breeds. Faster growing, healthier progeny are obtained that produce lean carcases. This breed therefore plays an important role in the beef production in hot and humid areas. It has now spread to over 60 countries in the world and both Australia and South Africa have their own Breed Societies.

Santa Gertrudis

The King Ranch, Kingsville, Texas, U.S.A. is located on land that was originally granted from the Crown of Spain by the Santa Gertrudis land grant. The new breed of beef cattle developed on the King Ranch between 1910 and 1940 was named after this Spanish land grant.

At the beginning of the 20th century, cattle from India were creating a lot of interest in East Texas. Although these animals were hard to handle and lacked the beef quality of the European breeds, they did thrive in the hot, dry Texas climate.

Richard King's
Cattle Brand

Registered 9th Feb 1869

On the King Ranch at this time were purebred herds of Herefords and Shorthorn, which by 1915 were the largest in the U.S.A. However, in 1910, the ranch was given a huge, black half-bred Brahman-Shorthorn bull. As an experiment, the manager, Robert J. Kleberg Senior, put this bull together with purebred Shorthorn bulls at pasture with four thousand purebred Shorthorn cows. The female progeny from this crossbred bull were mated to Shorthorn bulls and all the male progeny were castrated except for a red bull calf called 'Chemera'. After a number of breeding seasons, it was obvious that the crossing of Brahman blood with the Shorthorn stock had produced the best range cattle for hardiness, size and ability to fatten yet seen on the ranch.

In 1920, after ten years of experimenting, the King Ranch produced a cherry red bull that they called 'Monkey'. He was by a $\frac{7}{8}$ Brahman bull and out of a Shorthorn cow that carried about $\frac{1}{16}$ Brahman blood. Monkey was the foundation sire from whose progeny the Santa Gertrudis breed took shape. He was used from 1923 until his death in 1932 by which time he had bred over 150 sons that were good enough to use in the breeding programme.

The new breed was recognised by the United States Department of Agriculture in 1940 and it became the first new cattle breed developed anywhere for more than one hundred years.

The Santa Gertrudis today carry approximately $\frac{5}{8}$ Shorthorn blood and $\frac{3}{8}$ Brahman blood. From the Shorthorn, the breed gets its good carcase conformation and the rich red coat colour. Inherited from the Brahman is the resistance to heat and insect pests. The prominent hump of the Brahman has almost been eliminated.

Numbers of Santa Gertrudis cattle were built up steadily on the King Ranch, until by 1946 they had about 5,000 cows of breeding age and were using 1,500 purebred bulls. No females were sold but numbers of bulls were either sold or presented to interested ranchers and friends. Then, in 1950, the first annual auction sale of Santa Gertrudis bulls was organised. Two thousand people gathered in a circus tent to watch as the top price bull sold for $10,000. Two years later the highest price rose to $40,000.

A Breed Society called Santa Gertrudis Breeders International was formed in 1951 and published the first Herd Book in 1954. A standard of excellence and a classification scheme were adopted and animals that meet the requirements are branded with an S.

The breed can now be found all around the world in more than 45 countries and is proving adaptable to both tropical and temperate climates.

Brangus

The United States Department of Agriculture carried out much of the early work on Brangus. Brahman and Aberdeen Angus were crossbred at the Experiment Station at Jeanerette, Louisiana. The work began in 1932 and at the same time a few individual breeders in the U.S.A. and Canada were carrying out private breeding programmes. All were interested in establishing a black, polled breed that would thrive under adverse conditions and have the carcase quality of the Angus.

On July 29th, 1949 the pioneer breeders met in Vinita, Oklahoma and organised the American Brangus Breeders Association. This was renamed the International Brangus Breeders Association (IBBA) as there are now members in Mexico, Central America, Argentina, Southern Rhodesia and Australia as well as Canada and nearly every State of the U.S.A.

The method of breeding up to Brangus consists of 'enrolling' purebred Angus and Brahman stock with the Brangus Breeders Association. Their crossbred progeny can be 'certified' as $\frac{1}{2}$ bloods and by mating these to Brahman, a $\frac{3}{4}$ blood is produced as the Brahman fraction is always given first. By putting these $\frac{3}{4}$ bloods with pure Angus the Brangus is obtained with $\frac{3}{8}$ Brahman and $\frac{5}{8}$ Angus. These are the animals that can be 'registered' with the IBBA and if bred pure the offspring are also Brangus. The certified intermediate crosses are simply the tool to get to Brangus and are only rarely crossed back to each other.

Brangus cattle have short, straight black hair on a black pigmented skin that is loose and thin. The medium length head has a broad forehead and broad muzzle, and is polled. The body is long and has a straight back as the hump is not in evidence.

The quality beef produced from this breed has the well-marbled flesh inherited from the Angus. Calves are born easily and grow quickly on their mothers milk. For this reason, Brangus bulls are used on commercial herds of non-Brangus cows, and can safely be used on maiden heifers.

Droughtmaster

The Doughtmaster has been developed in Australia and for a relatively new breed it has a long history.

Bos indicus, or 'zebu' cattle were imported to Australia from India on the supply ships that provided for the early settlers. In 1843 and again in 1872 these animals were introduced to the Northern Territory. No true breeding programmes were begun until the turn of the century when Mr. D. Le Souef, curator of the Melbourne Zoological Gardens, thought that 'zebu' cross cattle would adapt well to North Australian conditions. The zoo sold a number of 'zebu' bulls to property owners in North Queensland about 1910.

Crossbreeding experiments were carried out using 'zebu' on different British breeds. The Council for Scientific and Industrial Research (C.S.I.R.) became involved and sent an officer to America to acquire 'zebu' cattle on behalf of the C.S.I.R. and private companies. Animals carrying Nellore, Guzerat and Gir blood arrived in Australia and were placed at four breeding centres in January 1934. These cattle and their descendants were kept under close observation for almost eight years before the C.S.I.R. would allow any to be sold. Then, in November 1941, the restrictions were lifted and property owners had a chance to buy and assess this new type of cattle. They proved popular as far more of the 'zebu' crosses survived the drought periods than did the British bred cattle.

By 1956 breeders had decided on the type of animal that they wished to establish and they named them Droughtmasters. The new breed was 50% Brahman, or American 'zebu', and 50% Shorthorn. Red was the preferred colour and the cattle could be either horned or polled. They had a small hump and the rounded rump that is a characteristic of the American 'zebu'.

Droughtmaster cattle grow well to produce beef under harsh range conditions of heat and poor pasture, but they also respond to improved conditions such as are found in feed lots. They have a high resistance to ticks and therefore suffer less from tick borne diseases.

The Droughtmaster Stud Breeders Society that was formed in 1956 has a breeding and classification programme designed to improve the breed. Special emphasis is placed on fertility, mothering ability, growth rate, docility and tick resistance. Herds have been started all over Australia and the breed has also been established in New Guinea.

Murray Grey

The name Murray Grey comes from the Upper Murray River Valley in Australia, where the first herd of grey cattle was started. In 1905, a light roan Shorthorn cow at Thologolong Station near Wodonga, Victoria, gave birth to a dun-grey calf sired by an Aberdeen Angus. The same cow went on to produce another twelve calves all of grey colour. These freaks of nature were kept for their curiosity value until it was found that they combined some of the best features of the Shorthorn and Angus breeds. Also, the grey colouration was dominant and a high percentage of grey cattle were produced even when crossed repeatedly with black Aberdeen Angus. Mr. Gadd, the owner of the herd, separated out the grey beasts and these formed the foundation of the new breed. Eventually, the breed obtained

Khillari (India)

Tharparkar (Pakistan)

Red Sindhi (Pakistan)

Texas Longhorn Bull (U.S.A.)

Texas Longhorn Cow (U.S.A.)

Brahman (U.S.A.)

Santa Gertrudis (U.S.A.)

Brangus (U.S.A.)

recognition and the Australian Murray Grey Beef Cattle Society was formed in 1962.

The high proportion of lean meat in the carcase has been one of the factors contributing to the success of the breed. Three carcases were shipped from Australia to take part in the U.K. Smithfield Show of 1967. They won first, second and third prizes in the Commonwealth carcase competition.

Murray Greys vary in colour from silver grey to dark grey and the thick skin is darkly pigmented. They are polled animals, early maturing, hardy and docile. Adult males weigh from 770–910 kg (1,700–2,000 lb) and females from 500–745 kg (1,100–1,300 lb).

The breed has been exported to other countries and is now established in the U.S.A., Great Britain, New Zealand and South Africa.

The Tasmanian Grey in Australia is another very similar breed. It has been developed from a crossbred calf born to a white Shorthorn cow in 1938 by an Aberdeen Angus bull. This grey calf bred at Parknook in Tasmania was the first in another line of grey animals.

Although a Breed Society has been formed, these cattle are almost identical to the Murray Greys.

Jamaica Hope

The Jamaican Government bought Hope Farm in 1910 in order to test different breeds of European cattle in a tropical environment. At the same time, their crossbred progeny out of native animals would be assessed. A dairy herd was established, and consisted of Ayrshire, Guernsey, Jersey, Brown Swiss, Holstein Friesian and Red Poll cattle.

Between 1910 and 1923, there was a development period in which pastures were improved and intensive work carried out on the control of tick-borne diseases. It was during this period that *Bos indicus* blood was introduced to the herd in the form of a Sahiwal bull from Pusa, India. He was mated to the cows of the different breeds in order to produce crossbred bulls that would transmit the heat tolerance factor present in *Bos indicus*.

At intervals in the breeding programme, certain of the European breeds were discarded as they did not appear to be contributing to the overall improvement. Finally, only the Jersey remained as it adapted better to the local conditions.

Native cows with *Bos indicus* blood were graded up with purebred Jersey and then bred to Sahiwal or Sahiwal cross bulls. After selection for production, only four of these grading up families survived at Hope Farm. One of these, the Narbrook family, was outstanding. Using this family as the basic material it was planned to develop them into a distinct breed.

In 1950 the Hope Farm herd was moved to the Bodles Animal Research

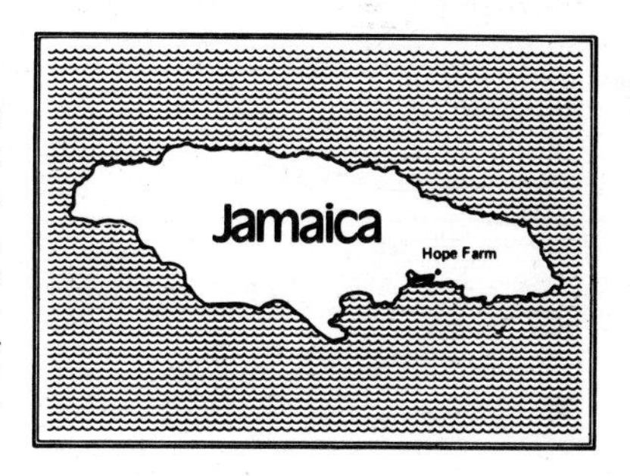

Station and bred as a 'closed' herd with no outside blood being introduced. The herd consisted of 177 cows and heifers, 70 calves and 40 bulls.

On June 25th, 1952 a ceremony was held at which the Hope Jerseys were formally declared a breed. They were called Jamaica Hope after the country and the farm where most of the early work had been carried out. In addition to the animals at Bodles, all those of a similar type on private farms were also elevated to breed status. The composition of the new breed was roughly 80% Jersey, 15% Sahiwal and 5% Holstein Friesian blood.

As most of the selection over the years had been on production and not on type or colour, then the breed lacks uniformity. However, the animals are basically of Jersey type, although they do not have the neat udders and level topline of the typical Jersey. Horns are longer and tend to turn upwards.

Luing

The Isle of Luing lies 15 miles off the west coast of Scotland. In 1947 the three Cadzow brothers, Denis, Ralph and Shane, purchased 2,000 acres of the 3,800 acre island and turned it into the home of the newest British beef breed.

The Cadzows did not set out to create a breed, but it happened as they were looking for cows that would produce a calf a year that could be taken to their east of Scotland farms for finishing. In their search for a cow with good mothering ability, hardy, long-lived and a good converter of roughage, they looked to the Beef Shorthorn crossed with the West Highland. These animals stood up to the wet west coast climate, were good foragers and the cows made excellent mothers. Fifty heifers of this cross were taken to the Isle of Luing and mated with a Beef Shorthorn bull.

Up to this stage, the Cadzow brothers used a traditional breeding system, but the problem came when they wanted to replace their crossbred females with cows of the same quality. They either had to buy them in and risk disease transfer, or they had to keep their own purebred Beef Shorthorn and West Highland herds in order to produce their own crossbreds. A decision was taken to try and fix the percentage of Shorthorn and Highland blood and use crossbred bulls.

A licence was obtained from the Scottish Department of Agriculture and the first crossbred bull selected for use was 'Luing Mist'. His sire was a top quality Beef Shorthorn 'Cruggleton Alister' and his dam was a Shorthorn-Highland cross. 'Mist' was mated to his half-sisters by 'Alister' and produced long bodied animals with a level topline and good legs and feet.

From 1949 to 1965, a programme of breeding and selection was followed in order to produce the type of animal for which the Cadzows were looking. In May 1965 they held an Open Day to let fellow farmers and breeders inspect their work. Most were impressed and in October 1965 there was an investigation by a committee representing all British breeds of cattle. The Luing was officially approved and in 1966 the Luing Cattle Society was formed and a Herd Book started. Two of the new Society's rules were unusual. Competitive showing was banned, although it was permissible to hold demonstrations, and no bulls sold at official sales were to be less than twenty months.

A further advance was made after 1969 when it was decided that more height was needed in the breed. Minimum standards were set and over the next four years the height was increased by 1·3 cm (0·5 in). Wither heights of mature males are now from 135–145 cm (53–57 in) and the average weight is about one ton (1,016 kg).

The colour of this new beef breed varies as no cattle are disqualified on colour. There are red, yellow and white whole colours as well as different roans and brindles. No horns are usually seen on Luing cattle because they are all dehorned at birth.

Luing cattle have been exported to Canada and New Zealand and both countries have started Breed Societies.

Beefalo

A hybrid breed that combines the merits of the American Bison, French Charolais and British Hereford has been called the Beefalo. Cattle breeders have been crossbreeding the 'buffalo', or American Bison, with domestic cattle since the end of the last century. Until recently, the progeny of these matings were sterile but a breakthrough came about 1960. A Californian rancher, Mr. D. C. Basolo found the ideal mix for the breed was $\frac{3}{8}$ Bison, $\frac{3}{8}$ Charolais and $\frac{1}{4}$ Hereford.

The Beefalo is an economical converter of roughage, such as grass or silage, into meat protein. It is claimed that ten month old animals can be ready for the butcher having reached 454 kg (1,000 lb) liveweight.

Beefalo vary in colour but are usually a shade of brown. However, some animals are black with the white head from the Hereford blood. They are long, deep, muscular animals but in spite of their mature size the calves only weigh from 18–30 kg (40–65 lb) and cows do not normally require assistance when calving.

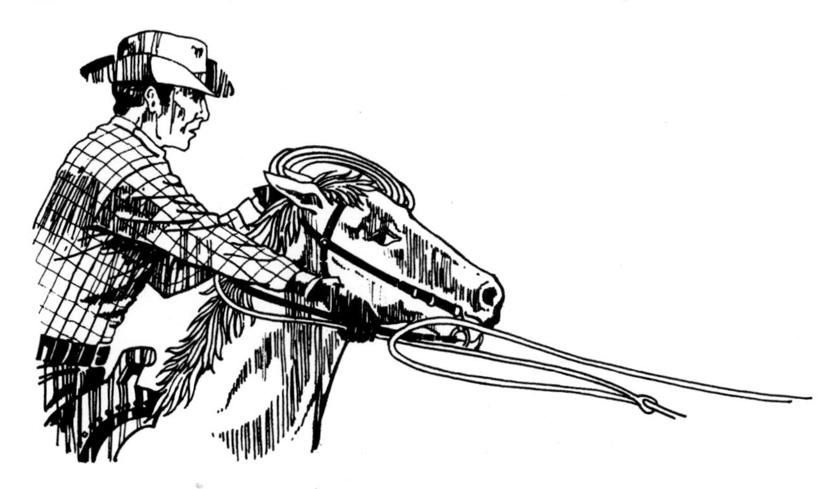

5 MILK AND MILK PRODUCTS

Historical Notes

Milk has been an important food item for over 6,000 years and is mentioned in the Sanskrit writings of ancient India. Artists impressions of cows being milked are to be found in temples and tombs of the early Egyptian and Mesopotamian civilisations. There is also evidence that from very early times, milk was converted into other products that could be stored more easily and kept for longer periods. In hot climates, with no refrigeration, milk goes sour within 24 hours.

In the Sumerian city of Ur, about 2,500 B.C. the people worshipped the Moon God and paid their dues to the temple priests in the form of pots of clarified butter and rounds of cheese. Butter was also put to good use some thousands of years ago by the Cretans. They allowed it to go rancid and then applied it to themselves as a cosmetic.

The Vikings took cheese with them when they raided the shores of the northern European countries and sailed across the Atlantic. Genghiz Khan's Mongol warriors, who conquered Asia and a large part of Europe in the 13th century, carried dried milk as part of their ration.

When the first European settlers landed in North America, they did not have cattle with them and many died because of the lack of suitable food, particularly milk. Later expeditions carried one cow and two goats for every six people.

Milk

Milk is the most complete single food that is found naturally; it contains sufficient nutrients to fully support the fast-growing young of mammals. It is, therefore, surprising to learn that cows milk is about 87% water. The 'dry matter' is made up of the following three main constituents: 3·8% fat, 3·4% protein and 4·8% lactose (milk sugar). Milk also contains a number of vitamins and minerals of which calcium and phosphorous are particularly important for living creatures. Unfortunately, there is an inadequate amount of iron in milk and certain vitamins are not present.

All over the world milk is drunk as part of the human diet. In the U.K. 60% of the milk produced is used for liquid consumption and the rest is manufactured into dairy products. This is a very high figure and few other countries have as high a liquid market. Most of the milk produced world-wide is used to make other products, such as butter, cream, yoghurt and especially cheese.

Cheese

Cheese can be made from the milk of goats, ewes, asses, mares, reindeer and buffalo, but most comes from cows milk.

The main aim of cheese-making is to take all the goodness from the milk and leave the water behind. This is achieved by adding rennet to warm milk, which then separates into curds. These are kept and the whey is drained off. Rennet is found in the gastric juices and stomach linings of many types of animals. It is thought that early man discovered the power of rennet to clot milk when he carried milk around in bags made from the stomachs of newly slaughtered animals. Today most rennets are prepared from the abomasum, or fourth stomach compartment, of unweaned calves. This is found to be the best source, although some plant extracts are also suitable.

Rennet consists almost entirely of the two enzymes rennin and pepsin. Rennin is responsible for precipitating the curd and pepsin turns the curd into cheese. The quantity of rennet, the quality and temperature of the milk and the temperature of the air are all important factors responsible for producing various types of curds. This, together with different techniques for making curd into cheese, results in a great varietyof the end product. These can be classified as either hard or soft cheese, with Cheddar at one end of the scale being very hard and Cottage cheese very soft, at the other extreme.

Individual farmers made all the cheese on their farms until recent times when the process became industrialised. Now most of the cheese produced around the world comes from factories.

The first cheese factory was built in the U.S.A. in 1851 in Oneida County, New York; others followed and by 1869 two-thirds of the cheese made in the U.S.A. was factory produced. An export trade to Europe built up and received a boost in 1860 when cattle disease caused a shortage of milk in Great Britain. English farmers quickly saw that to compete with their American counterparts they would need to mechanise cheese making. In 1870, the first factory was opened at Derby and used the milk from 300 cows on 13 farms.

From its primitive beginnings, the cheese industry has grown until today there are thousands of people employed in large cheese factories in all the industrial nations of the world.

Butter

Butter has had many uses down through the ages. In Northern Europe in the middle ages it became an important part of the human diet. The cooler climatic conditions there were more favourable for its preparation and storage. By the 12th century, butter was being exported from the Scandinavian countries. Up until the middle of the 19th century, butter making was a farm industry. The women used to make it in the summer months when there was plenty of spare milk produced by their cows from the summer grass. As there was a good supply of butter throughout the summer, it was sold cheaply and then in the winter became expensive and almost unobtainable.

Butter is not made from whole milk but from cream. An early method of obtaining the cream was to skim it off the top of the milk that had been allowed to 'set' in shallow pans in a cool place. The cream rose to the top of the milk and had to be skimmed off before the milk turned sour or the butter produced had an unpleasant flavour. The disadvantage of this process was that it took from 24 to 36 hours. A by-product was the so-called 'skim milk' that could be fed to farm animals. Towards the end of the 19th century, a separator was invented that used centrifugal force to separate the lighter cream from the heavier 'separated milk'. At first, this separator was worked

Well Kno
Milk Pro

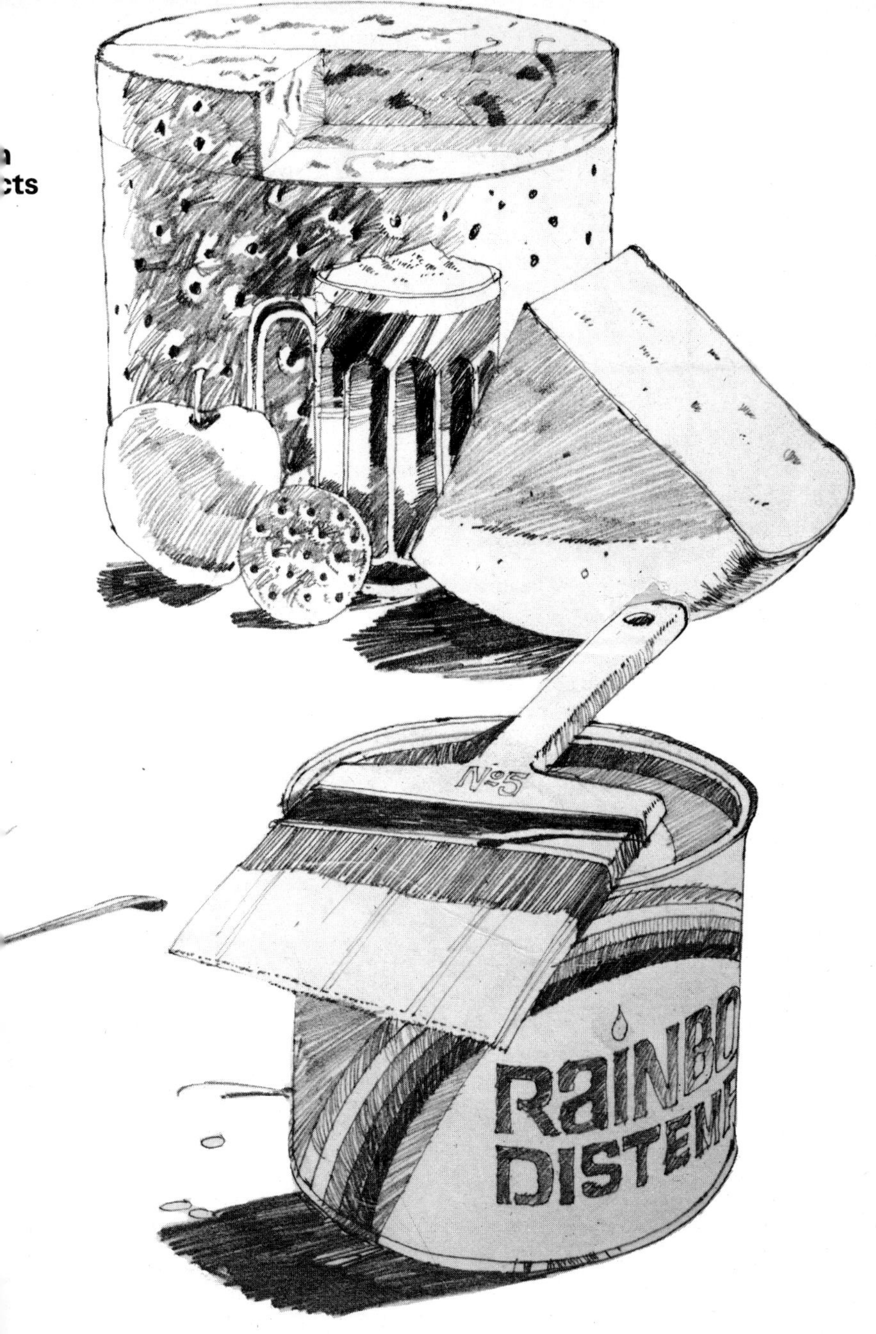
№5
RAINBO
DISTEM

by a handle, but later models were motorised and could separate up to 4000 litres (880 gallons) of milk in an hour.

Cream is highly concentrated milk which is rich in fat and is sold as a luxury food item. In butter making, the fat content of the cream is separated from everything else. Each little mass of fat in milk exists as a minute independent globule. In order to get these globules of fat to coalesce a process called 'churning' is used. A good deal of hard labour was involved in revolving the wooden churns by hand before butter making became mechanised.

The first factory to produce butter was built in the U.S.A. in 1861 in Orange County, New York and began by accepting whole milk from the surrounding farms and separating it into cream and skim milk. Later, other factories were built that only took cream from the farmers who installed separators on their farms. Both systems are still in operation around the world today, although the best butter tends to be made from factory separated cream.

There are a number of distinct processes in butter making, of which the three main ones are *neutralisation*, *pasteurisation* and *churning*. If a factory accepts daily delivery of fresh milk or fresh cream then there is normally no need for the neutralisation stage. However, if sour cream is to be made into butter then its acidity has to be reduced or the cream might curdle on heating and the resultant butter go bad. In order to lower the acidity, neutralising compounds, such as sodium bicarbonate, are dissolved in water and added to the cream by means of a steam injector or pump. During neutralisation, the cream should be at a temperature of about 32°C (90°F) and constantly agitated.

In many countries, there is a law which insists that all milk and cream for butter making is pasteurised. The object of pasteurisation is to destroy any harmful bacteria, to enhance the keeping quality of the butter and to produce a uniform flavour in the end product.

The process of pasteurisation involves heating cream with steam to reach the required temperature which is usually around 85°C (185°F). After this heat treatment, the cream is cooled and kept in vats for some hours to allow the fats to crystallise. This procedure is necessary to ensure good quality butter.

The cream is allowed to run from the vats into large metal churns until they are just half full. This is critical, as during churning the cream swells and may completely fill the churn. The churn is then revolved at speed which causes the fat globules to coalesce into granules. When these butter granules are about the size of peas the churning is complete. This normally takes between 30 and 40 minutes. The remaining liquid called buttermilk is then drawn off through a fine strainer in order to catch any granules. As soon as this is done the surface of the butter is washed in a spray of water. Sometimes, the butter is salted after washing in order to cater for the demands of certain markets.

Cultured Milks

Cultured milks are made from fresh milk which is allowed to go sour in a controlled way, using certain bacteria which convert the milk sugar, lactose, into lactic acid. In hot countries, these cultured milks have been made for hundreds of years, as the valuable

Droughtmaster (Australia)

Murray Grey (Australia)

Jamaica Hope Cow (Jamaica)

Jamaica Hope Bull (Jamaica)

Luing (Scotland)

Beefalo (U.S.A.)

Kerry (Ireland)

Dexter (Ireland)

nutrients in the sour milk are preserved in a palatable way by the lactic acid.

Some of the cultured milks are alcoholic, as a lactose fermenting yeast is introduced during the preparation. The most famous of these is *Kefir* which is thought to have originally been prepared in the Caucasus mountains in Russia. In the U.S.A., two non-alcoholic cultured milks called Buttermilk and Acidophilus milk are widely consumed, whilst in Europe one of the most popular types is Yoghurt.

The bacteria which gives Yoghurt its individual flavour was first isolated and named by the French Nobel Prize winner Dr. Ilya Metchnikoff. He was working at the Pasteur Institute at the beginning of this century when he made his discovery and called the organism *Lactobacillus bulgaricus.* Dr Metchnikoff thought that these bacteria could somehow be responsible for long life and good health because Yoghurt was an important part of the diet in Bulgaria where the people were known for their longevity. Although his theory was never proved, it did start a world-wide interest in Yoghurt as a health food. In many countries Yoghurt and other cultured milks are traditionally prepared in the home. However, in countries with large dairy industries these milk products are produced under scientifically controlled factory conditions.

Casein

Casein is one of three major proteins in milk and is unique as there is no other similar protein found naturally. It is one of the chief constituents of cheese and Cheddar contains some 33% of casein.

It is possible to precipitate casein from milk by the use of dilute acids or rennet. Commercial casein is made from skim milk after the cream has been removed. In its pure state casein is white, odourless and tasteless which makes it an ideal additive to foods in order to increase the protein content.

Casein has been used in many different ways, but has now largely been replaced by other products. In the plastics industry, casein was manufactured into buttons, umbrella handles and many other useful articles but now modern plastics have taken over. Casein glue was widely used at one time, as were casein bristles in brushes, and casein for the sizing of paper to make it less absorbent to printers inks. The leather industry uses casein as an ingredient of seasonings and pigments, and it has had a long history in the manufacture of paints. Casein paints were used in ancient Rome and even now artists use them to create an 'oil painting' effect. The advantages are that casein paints dry quickly and can be used with bristle brushes. They are water resistant when dry but too brittle for canvas so must be used on rigid boards. Some emulsion paints and distempers for house decorating also contain casein.

Butter and Cheesemaking

MAELOR CREAMERY

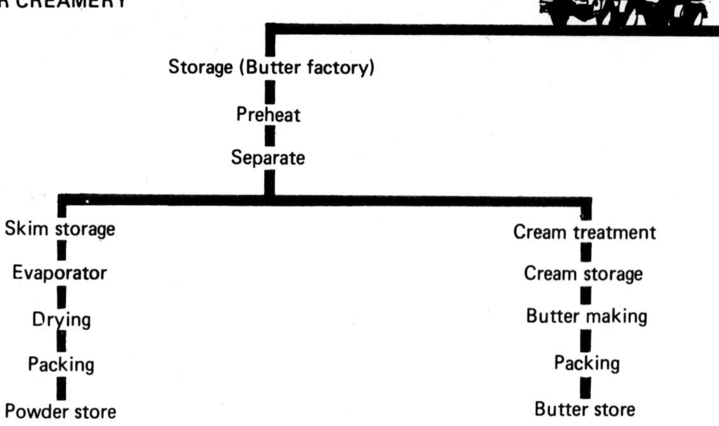

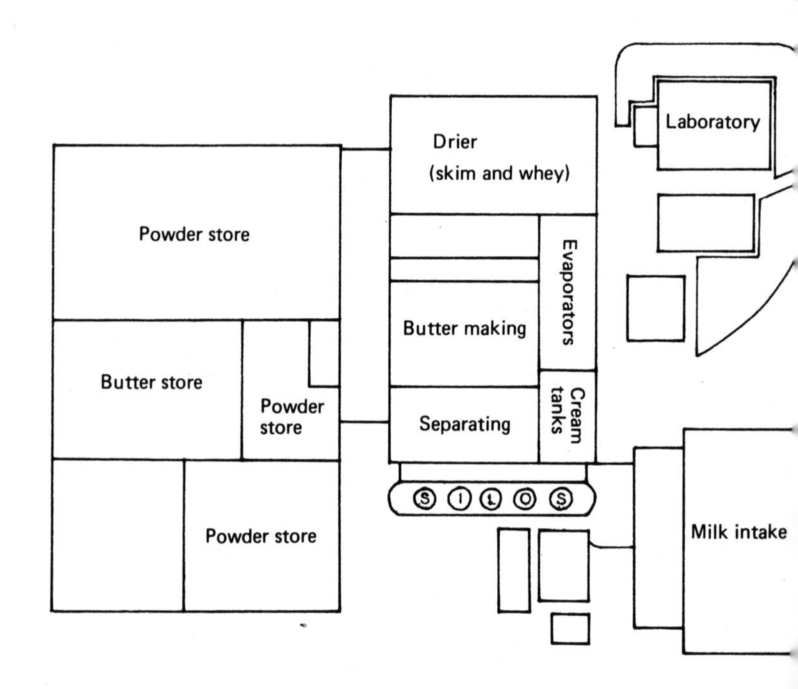

MILK

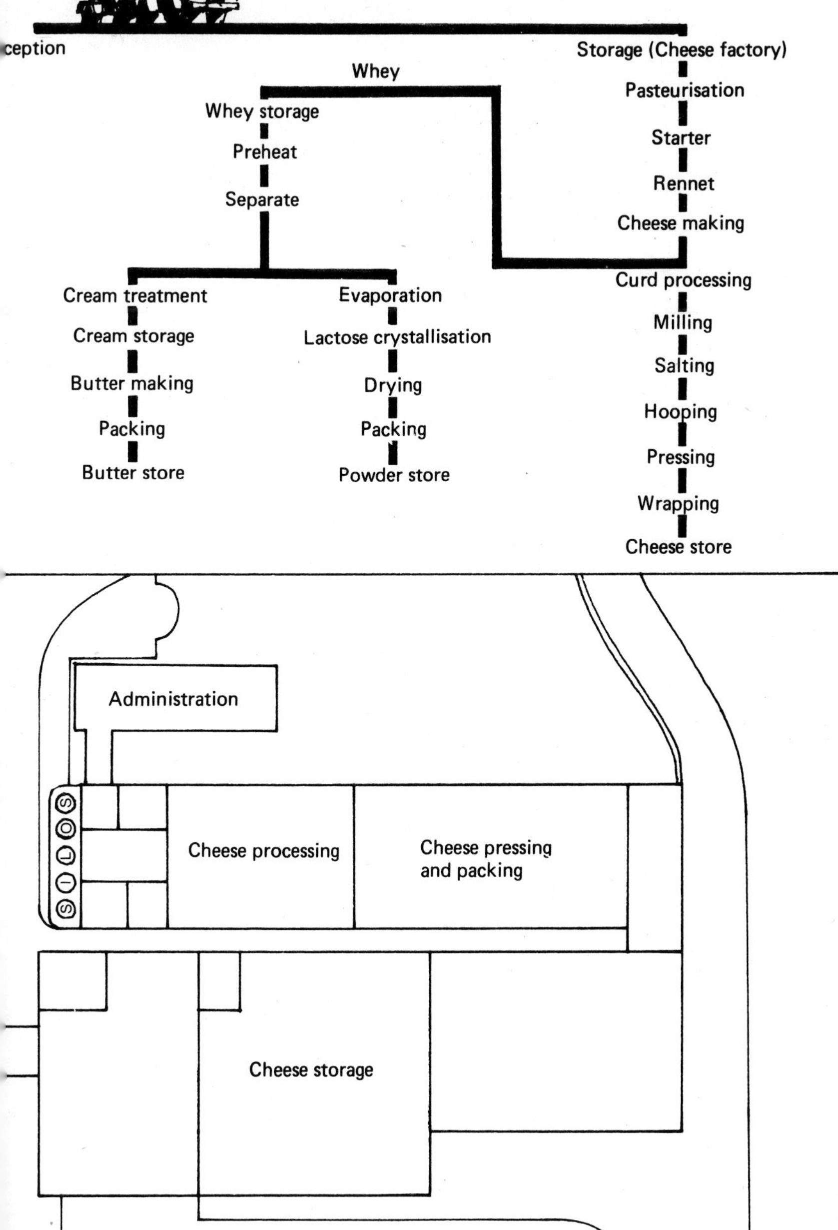
ception
Storage (Cheese factory)
Whey
Pasteurisation
Whey storage
Starter
Preheat
Rennet
Separate
Cheese making
Curd processing
Cream treatment
Evaporation
Milling
Cream storage
Lactose crystallisation
Salting
Butter making
Drying
Hooping
Packing
Packing
Pressing
Butter store
Powder store
Wrapping
Cheese store
Administration
SILOS
Cheese processing
Cheese pressing
and packing
Cheese storage

6 MODERN BREEDING TECHNIQUES

Cattle breeders are always seeking new ways of improving their stock and one of the quickest ways to improve a cattle population is to allow only the very best animals to produce the next generation. Down through the centuries, the poorest cows in herds have been culled and the better ones mated to a superior bull. However, there was always a limit to the number of calves a good cow could produce and to the number of cows a herd sire could serve. Recently, these limits have been dramatically broken by the discovery that bull semen can be frozen and kept alive for an indefinite period, and that cows can be induced to shed many ova, (eggs that have been fertilised by sperm), which can be transplanted into host animals. These techniques, which will be described later in this chapter, can be used to breed thousands of progeny from any one bull and a greatly increased number of offspring from any one cow.

Artificial Insemination

In order to appreciate the significance of frozen semen in a breeding programme, it is necessary to briefly trace the history of artificial insemination, often abbreviated as AI. The Arabs are credited with inventing the technique and applying it to the breeding of horses. A report dated 1322 tells of a successful insemination of a mare by semen collected from an Arab stallion. However, it was not until 1899 that a Russian veterinarian developed the practice in sheep and cattle and began to train people to artificially inseminate these animals in the years leading up to World War I.

Whereas in natural service a bull is allowed to mate normally with a cow, in artificial insemination the process is humanly controlled. The bull is stimulated and produces semen which is collected by man and then used to impregnate cows that show signs of being ready for inseminating. Good fertility results are obtained if the inseminator places the semen in the correct spot in the cow's uterus, so that the spermatozoa can travel up each uterine horn towards the site of fertilisation as the egg is released from the ovary.

In Russia, a sponge or swab was used when making the early attempts at collecting bull semen and no real progress was made until the development of artificial vaginas in Italy and Russia. As there was very little advantage in putting the pure semen back into cows, a diluent was produced which was used to reduce the density of the semen to about 12,000,000 spermatozoa per millilitre. When carrying out an artificial insemination, a millilitre (1 ml) was the amount of diluted semen used.

Fresh, diluted semen could be kept for several days before fertility dropped to an unacceptable level. Then the unused semen had to be discarded and another collection taken from the bull. This was a wasteful process, even though it enabled a good bull to sire far more progeny than under natural service conditions. With the discovery that frozen semen would keep alive for unlimited periods came the possibility of artificially inseminating some 100,000 or more cows a year with semen from the same bull.

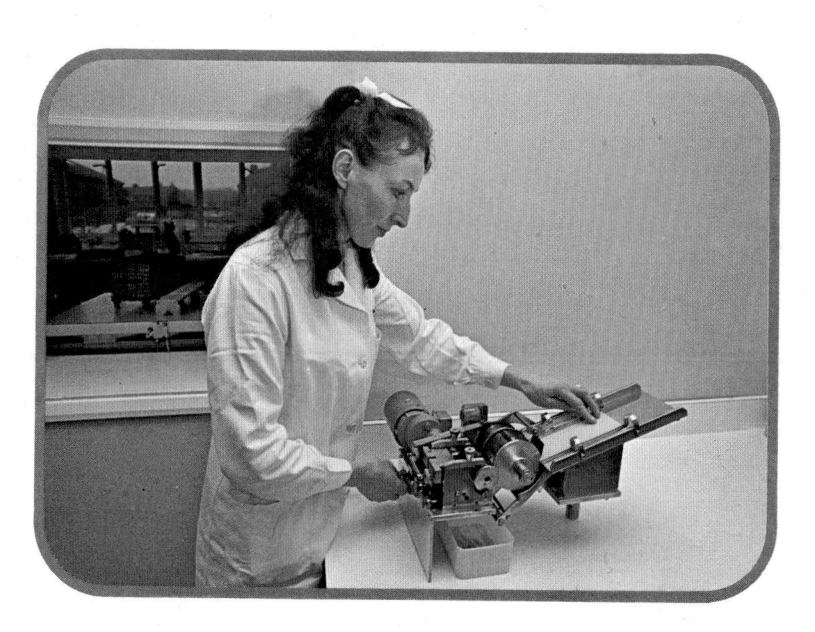

Labelling straws prior to filling with semen

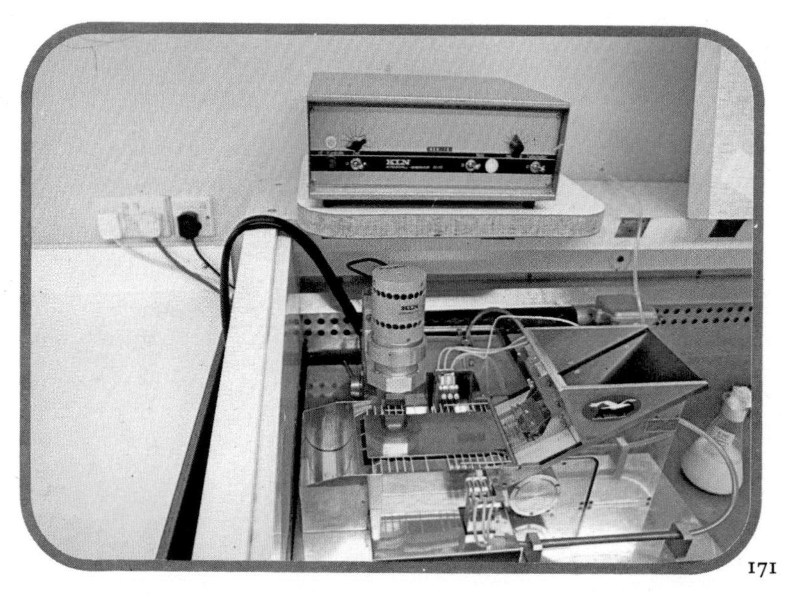

Filling straws with semen before freezing

Labelled straws of stored semer

ALSOPDALE SUNBEAM 2ND
MONKLAND CHANCELLOR
BOYEYLONELY 1ST
PRINCE PROPEL OF ADDINGTON
MMB ISENO
MONKLAND CHANCELLOR
MMB HERMANN
MMB IMP74 OSSO
ZEEBARN CARINTHIAS CARDINAL
CASTLEMARTIN BLACK GORT
MMB APOLLON
FANFARON
PRINCE PROPEL OF ADDINGTON
D102
PROSPERO OF QUEENS LETCH
PROSPERO OF QUEENS LETCH
MMB APOLLON
PROSPERO OF QUEENS LETCH
PRINCE PROPEL OF ADDINGTON
ORGREAVE VICTORY
ELBRIDGE KEYSTONE 3
FANFARON
PRINCE PROPEL OF ADDINGTON
ACTHORPE PEGASUS
MMB IMP74 OSSO
ORGREAVE VICTORY
POLWHELE TRUMPETER 2ND
PRINCE PROPEL OF ADDINGTON
ORGREAVE VICTORY
MMB REBHOLZ
CARINTHIAS CARDINAL
ELBRIDGE KEYSTONE 3
ORGREAVE VICTORY
TOWNHEAD WESTMINSTER
ALSOPDALE SUNBEAM 2ND
FANFARON
CASTLEMARTIN BLACK GORT

Straws of semen about to be placed in liquid nitrogen for freezing

Milk Marketing Board Central Storage Unit for semen at Towcester

Frozen Semen

In England, in 1948, it was found that semen could be frozen solid at a temperature of —79°C (—112°F). During 1951–52 trials were carried out using semen that had been frozen and stored in dry ice and alcohol. The first success came when 38 cows were inseminated with the semen after it had been thawed out and 30 of the cows became pregnant. Shortly afterwards the Americans developed a technique of freezing bull semen in liquid nitrogen at —195°C (—320°F). Originally, the semen was enclosed in glass ampules but in 1967 the French used lengths of small bore plastic tubes called 'straws' that held 1 ml of semen. These straws proved a very effective way of freezing and storing semen. In some countries, straws have been reduced in size and hold only 0·25 ml of diluted semen. Straws can be printed with the bull's name, herd book number and other identification numbers before being automatically filled with semen. Also, it is possible to use different coloured plastic for each breed as a method of coding the straws. One other alternative method of storing frozen semen is in pellets. Drops of diluted semen are pipetted into small indentations on a block of solid carbon dioxide. On contact, these freeze into pellets and can be wrapped in foil that has been coded with the bull's particulars. Before insemination, the frozen pellet is reconstituted in a sodium citrate solution and is ready for immediate use.

By freezing, semen can be stored for years and many organisations specialising in AI have large storage units. These units have many stainless steel storage flasks containing liquid nitrogen to keep the semen at very low temperatures. All storage units have high standards of hygiene, as harmful bacteria and viruses can also be preserved in a frozen state.

Advantages of Artificial Insemination

One of the most obvious advantages of AI is that many breeders may use the same bull in their herds simultaneously. All then gain if the bull produces outstanding progeny. However, if the bull breeds inferior stock, then a large number of breeding plans would be spoilt. Fortunately, artificial insemination can be used to prove whether or not a bull will produce satisfactory offspring. The method is called a 'Progeny Test' and can be used for either beef or dairy bulls. Semen is collected from a young bull and used on a limited number of cows in many herds. The progeny are evaluated for their beef or dairy potential, depending on the breed, and the results are used to rate the bulls. The best bulls are kept to sire large numbers of progeny by artificial insemination whilst the poorer bulls are culled.

Bulls can transmit diseases to the cows. By making sure that their sires are free from all known diseases, the AI organisations are reducing the likelihood of spreading disease via the semen. In the U.K. bull semen is stored in liquid nitrogen at special quarantine units for at least 28 days before being used. This is to guard against the donor bull carrying a disease that was not detected at the time of the semen collection. If the bull subsequently falls ill, then it is possible to discard the semen before any is used.

All technicians carrying out insemina-

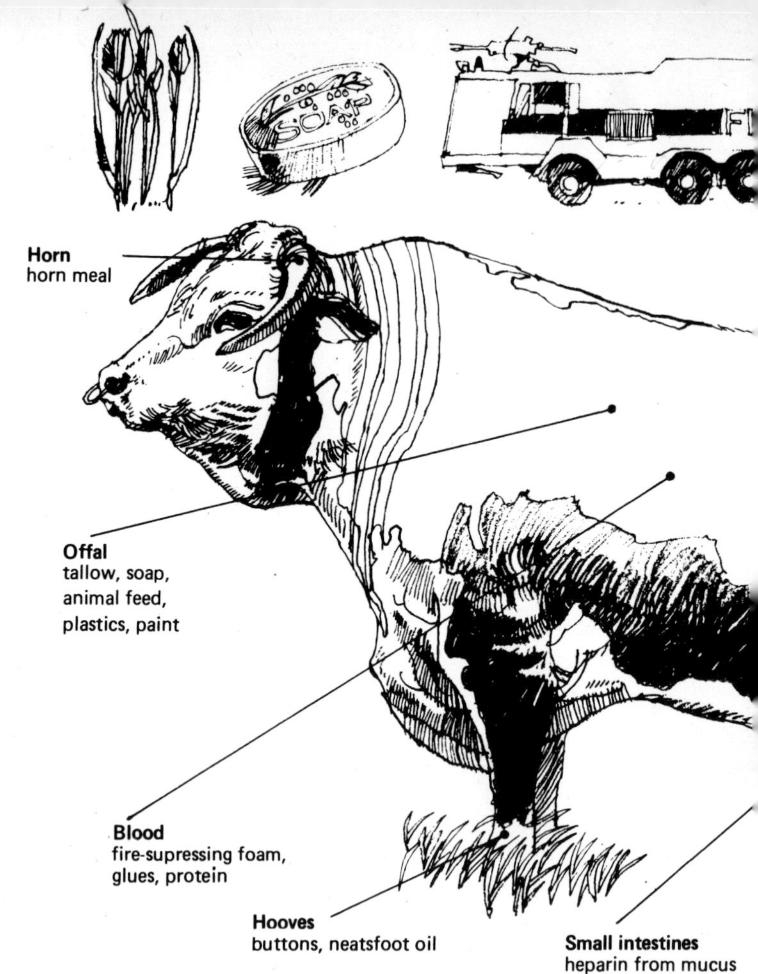

'The Fifth Quarter'

When cattle are slaughtered, nothing is wasted. Valuable by-products are obtained from the 'fifth quarter', or remainder of the animal after the meat has been removed. It is said that the profit from the sale of these by-products, including leather, helps the meat industry to survive.

One of the by-products that has been used for centuries is tallow, which is produced by rendering down all the inedible waste, or offal. The early use of tallow was in the manufacture of candles but now it is used to produce laundry and kitchen soap. The U.S.A. is the largest producer of tallow, with an annual production of 1 million tons. Some 40% of this American output is incorporated into animal feeds.

Bones yield bone meal fertilisers, and glycerine, some of which is converted to glyceryl trinitrate for explosives

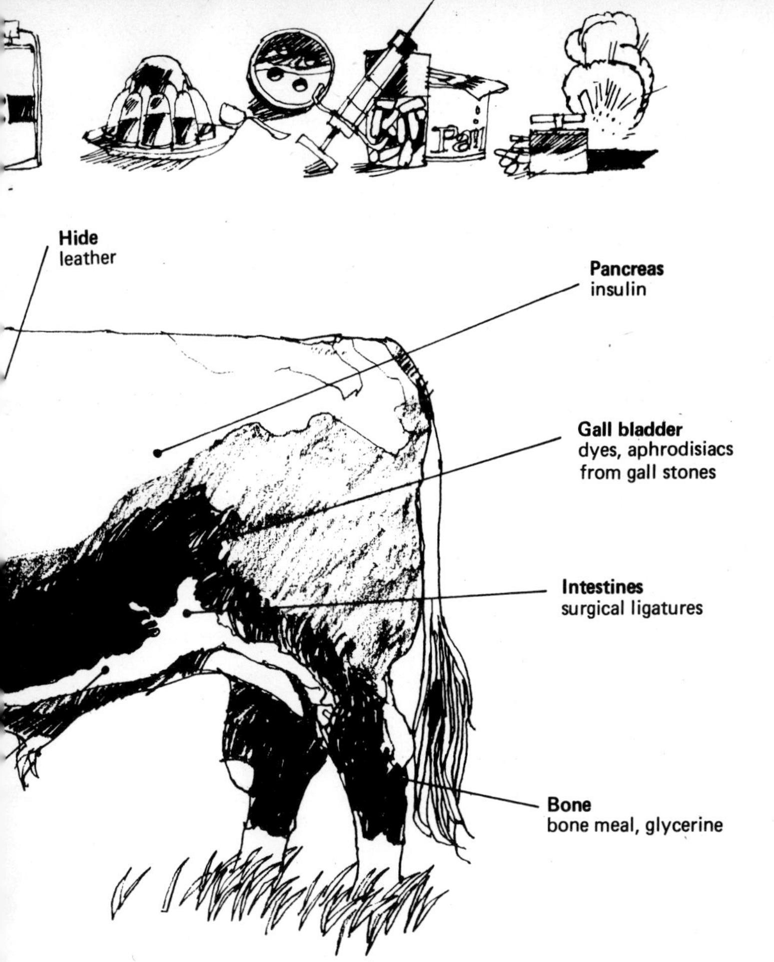

manufacture. Hooves are made into buttons, and neatsfoot oil. This oil was once in demand for preserving harnesses, but it is now mainly used as a cutting oil in engineering. Horns are ground into horn meal and at one time were shaped into knife handles. Proteins, glues and fire suppressing foam are made from blood. Dyes come from the gall bladder and gallstones are ground up and sold as an aphrodisiac to far eastern countries. The intestines are made into surgical ligatures, and heparin, which is used to treat blood clots, is extracted from the mucus of the small intestine. Pharmaceutical firms make good use of certain organs and glands.

The pancreas is particularly important as it yields insulin which is in great demand for the treatment of diabetics.

tions must wear protective clothing which is disinfected between farm visits. The straw of semen is taken from the liquid nitrogen container, allowed to thaw out before being placed in the inseminating gun. The insemination is performed using disposable equipment, which is discarded on completion of the service.

A further advantage of artificial insemination is that it can help the numerically smaller breeds. In any cattle population there is a danger of 'in-breeding' through the mating of close relatives, and the smaller the population the greater the danger. Inbreeding can produce inferior stock and is normally to be avoided. By collecting banks of semen from bulls of unrelated blood-lines, it is possible, through AI, to prevent any 'undesirable' matings.

There is always a danger of breed extinction where cattle populations are small. However, it is possible to reconstitute a breed providing there are frozen semen banks from bulls of that breed in existence. By using semen of the extinct breed on any cows, but preferably those with similar characteristics to the dead breed, the resulting progeny are 50% of the desired end product. These animals are then artificially bred to more of the frozen semen and after five or six generations the lost breed is back in existence. Thus AI can become a tool of the conservationists. Rare breeds that are in danger of extinction, may carry factors that could turn out to be commercially useful in years to come, so some organisations are actively banking semen from these breeds. In the future, this semen may be very valuable for use in crossbreeding programmes where a desirable character of an extinct breed is needed to combine with the characteristics of an existing breed to produce a new type of animal.

World Use of Artificial Insemination

In spite of all the advantages of AI only about 7% of the total cattle population of the world are mated this way. Many countries are not sufficiently well developed technically to make use of the technique. Only 1% of the 240 million cows in India are artificially inseminated in spite of serious attempts to build up a programme. This should be compared with Israel and Finland, who each put over 99% of their cows to AI. It is very easy to transport frozen semen from one country to another, but unless the receiving nation has trained technicians and a ready supply of liquid nitrogen in which to store the semen then an AI programme is virtually impossible.

A growing export trade is developing for frozen semen from the best bulls in North America and Europe. This semen is being used all around the world to breed better animals for milk and beef production. Native cattle in South America and Africa are being crossed with the best breeds from the agriculturally developed countries. These native animals will soon lose their separate identity as they are graded up to Devon, Hereford, Brahman and other breeds.

In Australia, where there were no native cattle, the breeders imported animals. Now, to improve their herds, they are importing semen of progeny tested bulls. At the same time, the present day Australians are testing the bulls born in their own county in AI breeding programmes. The same process

Australian Charolais heifer calves produced from semen imported from the U.K.

Simmental calves bred in Australia from imported semen

A successful transplant operation in Australia resulted in these 12 Limousin calves being born to recipient cows. All the fertilised eggs were recovered from the heifer in the background.

Crossbred calves in Kenya out of native cows bred to imported Devon semen

A crossbred calf produced in Australia by use of imported South Devon semen on a Shorthorn Cow

is going on in New Zealand and the two countries between them inseminate some two million cows. This is a long way behind Europe, where 64 million cows go to AI and North America with 11 million. However, the situation could change dramatically as the rest of the world catches up with these leaders.

Controlled Breeding

Where AI of cattle is practised, there is always the problem of the correct time to carry out the insemination. Some cows show clear signs of being ready for mating whilst others do not appear to exhibit any symptoms of being 'on heat'. This is the term used when a cow is at the right stage of her breeding cycle to be successfully served. Cows have a cycle that last approximately 21 days and during this time can be on heat for about 15 hours, although individual animals vary between 1 and 30 hours.

The human who has to take the decision to inseminate can sometimes be wrong, whereas a bull in the herd knows instinctively. In order to help the human decide, many aids to heat detection have been tried. One of the most successful being the running of a castrated bull with the herd. This animal can spot a cow on heat and, whilst attempting to serve her, can give the farmer visual evidence of the cows suitability to be artificially inseminated. However, not all breeders wish to go to the trouble and expense of having an extra animal on the farm.

In the last few years, scientific developments have taken place that have rendered the guessing, regarding cows being on heat, unnecessary. The synthesising of hormone-like substances called prostaglandins, and the discovery of the way in which they affect the breeding cycle, has led to controlled breeding where cows can be mated on predetermined dates.

Prostaglandins occur naturally in a wide variety of animal tissues and were first discovered by the Swedish scientist von Euler in 1934. There are many different prostaglandins and they have been divided by structure and function into four distinct groups. One of these groups, called the F Series, can be used to induce heat in cows that are at a certain stage in their breeding cycle. This particular stage lasts for about 12 days in mid-cycle, and if cows are injected then with an analogue of prostaglandin, they will come on heat 2 to 3 days later. However, some of the cows will not respond, as they are not in the correct stage of their cycle. To overcome this problem, and get a group of cattle all on heat at the same time, two injections are given with an eleven day interval between. After the first injection, on day 1, some animals will be brought on heat but not inseminated. On day 12, they can have their second injection and brought on heat again. The remaining animals that were in the wrong stage of the breeding cycle on day one will be at the right stage on day 12 when they are given their second injection. Therefore, all animals come on heat 2 to 3 days after the second injection. Whether or not these animals show visible signs of being on heat, they are artificially inseminated 72 hours after the second injection. This is the optimum time to achieve a satisfactory conception rate. Some owners of cattle arrange to have another insemination of the whole group 24 hours after the first insemination. This has been shown to improve the numbers of animals subsequently found in calf.

In this way, large groups of cows can be mated at one time without the worry of heat detection. Owners of dairy cattle find this technique valuable when mating heifers that are notoriously difficult to spot on heat. Before the advent of prostaglandin analogues, many would have been put in groups with a young bull. Now it is possible to mate them to top sires owned by artificial breeding organisations. Breeders of beef cattle are using the prostaglandin technique to inseminate groups of cows to beef bulls, in order to obtain uniform bunches of calves to put through their feeding and fattening processes. This is because most of the cows inseminated on the same day will calve within a two week period at the end of their nine month gestation. A bull could not serve a group of females on anyone day because they would be at different stages of their 21-day breeding cycle. As the calving dates would also be spread out there could be progeny of the bull born two months apart—a much less even group than that produced by using a combination of prostaglandin injections and AI.

Egg Transplants

The techniques so far discussed have involved making the best use of superior bulls through AI, deep freezing of semen and controlled breeding. Recently, all three innovations have been combined with other scientific developments in order to take advantage of the exceptionally good cow. Rapid improvement can be made in a cattle population by transferring fertilised eggs from the best cows to less valuable foster mothers. A single cow in her lifetime can, at best, give birth to some sixteen calves, and the average is much nearer four. However, the potential is enormous as the immature eggs present in the ovaries of a young cow can number 100,000. Normally only a fraction of these mature, and an even smaller proportion are fertilised and become living cattle.

An injection of certain hormones into cows, just after they have come on heat, causes numbers of eggs to mature and be released, rather than just one. This phenomenon is called *superovulation* and up to 80 eggs have been released at one time. After this treatment, the chosen donor cow is artificially inseminated a number of times in a 24 hour period—thus ensuring that there is a good supply of semen present to fertilise the eggs. A surgical operation is usually needed to collect the fertilised eggs for transplantation, although there are non-surgical techniques being developed. The surgery is carried out 5–7 days after the donor animal was on heat and between 50% and 75% of the fertilised eggs, or embryos, are recovered. A disadvantage is that this expensive operation is tedious and few cows can undergo it more than three times and still produce embryos for transfer.

The embryos are flushed from the donor cow in the course of the operation, using a simple solution of salts, sugars and protein. This solution can also be used as a medium in which to store the embryos between collection and transplantation to the uterus of the recipient cow. Ideally, the transplant should be carried out within a few hours of the embryos being collected in order to achieve acceptance by the foster mother. Also, the foster mother must be at the same stage in her breeding cycle as the donor cow. Originally, this presented problems unless a large number of possible foster mothers could be collected together. Now an analogue of prostaglandin is usually injected into the donor and recipient cows at the start of each transplanting scheme in order to get them all on heat together.

From an average of 15 super-ovulated eggs collected per operation about four progeny are eventually born, but this should improve in time.

Experiments have been carried out with deep frozen embryos. In 1973, a normal bull calf, 'Frosty', was born at the Animal Research Station, Cambridge, England after transplantation of a frozen and then thawed cow embryo. Success rates are not yet very high with frozen embryo transplants and the methods used are expensive. When the techniques are perfected it will be possible to fly potential herds of genetically superior cattle, from one side of the world to the other, as fertilised eggs in a flask of liquid nitrogen.

7 SHOWS AND SALES

Breeding for improvement has been mentioned many times throughout this book and there is no doubt that breeders get great personal satisfaction from producing better cattle. The pleasure is far greater if a breeder is able to prove to his friends and neighbours that his animals are the best in an area. What better way to do this than by entering cattle at nearby Agricultural Shows in the hope that they will win prizes!

Winning at local competitions encourages entering at national level and showing cattle has become an accepted practise in many countries. The breeders winning the prizes usually find that they obtain more money for their stock, when eventually sold, as prizewinning cattle are a good advertisement for any herd. Showing has, therefore, been linked to selling, and many shows and sales are held where the animals compete for prizes before being auctioned.

Shows

Centuries ago 'Fairs' were held in China where merchants could display and sell their merchandise. This practise was adopted in Europe and farmers were present among the various sellers of goods. Cattle, as well as farm products, were brought for sale. It is these fairs that were the forerunners of todays agricultural shows.

The first competitive shows for livestock were those held in Britain during the last years of the 18th century. They were sponsored by Agricultural Societies whose members were devoted to agricultural progress including the improvement of livestock. The Bath and West of England Society held the very first show in 1797, followed a year later by the Sussex Agricultural Society. Cash prizes were offered, and owners of the best bulls had to undertake to allow them to serve a certain number of cows belonging to Society members during the next twelve months, for an agreed fee. In this way, shows influenced the type of beef and dairy animals bred, as the Champion beasts would sire progeny on a number of different farms.

Early shows only attracted local breeders, because of the problems of primitive transport. When cattle were walked long distances to shows or markets they lost weight and condition. This meant that they did not look so smart in the show ring or make as much money in the sale ring. However, by the end of 1848 nearly 5,000 miles of railway were in operation in Britain. Farmers and cattle dealers quickly realised the advantages of this form of transport and by 1913 the railways were carrying 3,792,000 cattle in the year (see page 146).

The Fatstock Shows held in London by the Smithfield Cattle and Sheep Society, and the higher prices being paid for meat animals at Smithfield Market, attracted cattle from greater distances. Many came by train, but in more recent years the lorry has proved to be a cheaper substitute. To save the cost of hiring transport, many breeders have their own lorries to drive themselves and their cattle to the shows and sales.

Nowadays cattle can be moved quickly and efficiently over long distances both in and between countries by lorry, train or plane.

Judging

At the very early shows, cattle were just exhibited for the onlookers to admire. Later, competitive events were staged where the animals were paraded in front of a judge and placed in order on their looks. Livestock are now judged by the known experts in animal breeding. These are farmers and breeders who have made good reputations by producing superior animals that have pleased their purchasers. Butchers, also, have been invited to judge beef cattle and particularly carcase competitions.

All judges have a great responsibility, for the cattle they choose as winners will normally be accepted by the farming community as the type to breed.

Judging has traditionally been done in 'Classes', where like is compared with like. Each breed being judged is divided up into the different types, or classes, of stock. Dairy cattle are split into milking cows and dry cows of various ages. In this way, heifers in milk that are young and attractive are not judged against older cows that have had a number of calves and lost their youthfulness.

Cows in milk can also be entered in type and production classes, where points are awarded for the amount of milk given as well as the appearance of the animals. These classes are most important as they give a good indication of a cows worth in commercial terms. In the same way, there are classes for beef animals where the beasts are judged live and then as carcases before an overall winner is announced. An attractive steer in the show ring can sometimes prove to be a poor buy for the butcher because of excess fat. Consumers today demand lean meat and this the trade must provide, rather than the fat joints that were preferred by our ancestors.

At certain shows, there are weight recorded classes for beef cattle where the judges take into account the animals daily liveweight gain (d.l.w.g.). The beasts are weighed on the showground, their estimated birth weight is deducted from this figure and the result is divided

Beef animal about to be weighed

Supreme Champion Blonde d'Aquitaine

Display of Prize Winners in the Grand Ring of the Great Yorkshire Show, England, 1976

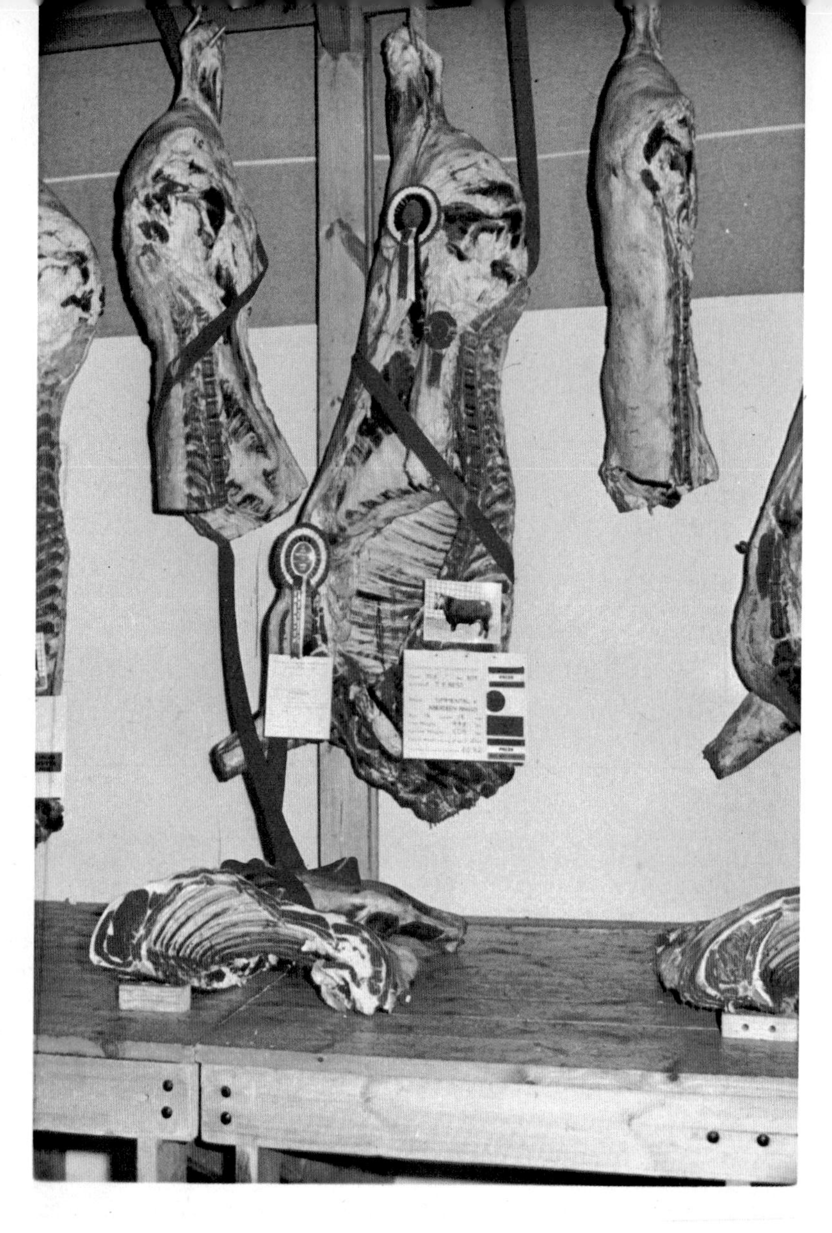

Supreme Champion Carcase at U.K. Royal Smithfield Show from a Simmental X Angus steer

by the number of days it has lived. This calculation produces the d.l.w.g. and the higher the figure the more points awarded by the judge to add to the score that the animal obtains for its looks in the showring. The combined points for weight and the type determine the winner.

Many producers of beef animals weigh them regularly in order to find out how fast they are growing. It is then possible to predict when they can be sold for meat and to decide whether they need fattening more quickly. Steers with a d.l.w.g. of only 0·91 kg (2 lb), on average, may make a loss as they are taking up valuable space and requiring costly labour for a long time before being sold. If they could be fed to grow at 1·13 kg (2·5 lb) per day, they could show a profit by reaching the required weight and going for sale sooner.

Although outstanding individual animals can be a credit to their breeders, far more prestige is gained from showing a group of animals of similar good breeding. At certain shows, there are classes for two or more daughters of a particular bull and also classes for cow families. As an advertisement, there can be nothing to surpass a line-up of good daughters of one sire, or a grand old cow parading with her offspring. It gives the onlooker confidence in the herds that produce this kind of cattle. Unfortunately, there are still prize-winning bulls that are sold for a lot of money and yet are never heard of again because they have produced inferior progeny. Also, there are good looking cows that are kept in order to win at shows rather than to milk at home because they are poor milk producers.

The Champions

All shows have their overall Champion animals that are top of all the first prize winners. Beautiful cows with perfect udders; bulls with deep bodies, level toplines and square rumps, and steers with meat in all the right places. Then there are Breed Champions picked from the winners of the classes and Reserve Breed Champions.

The Reserve animals are traditionally of the opposite sex to the Champion, unless the show consists entirely of bulls, or cows.

At the larger shows, there are more classes and therefore more Champions, because the entries tend to be split into Junior and Senior classes, depending on age. This means that there are Junior and Senior Champions in each breed competing for the award of Supreme Champion of the Show.

After the judging, there is usually a parade of prizewinners led by the Champions. All the animals wear the rosettes and ribbons that they have won. The different colours denote their position in the awards list. Sometimes, as at the U.K. Royal Show, there is a commentator telling the public about winning animals and the merits of their respective breeds. Finally, the animals line up by breed in the main arena to be admired by the spectators.

Sales

Cattle are a marketable commodity all around the world whether they are required for draught, meat or milk. Normally, pedigree cattle are sold separately from non-pedigree, or 'commercial' cattle, as they go to different groups of buyers. Pedigree cattle are bought and sold by breeders looking for new bloodlines and ways of improving their herds. Beef calves and some calves out of dairy cows are purchased for fattening by farmers. Some beef animals are then sold again at a halfway stage in the fattening process, as 'store' cattle, to other farmers who specialise in producing finished butcher's beasts. The rest are taken right through to eventual sale at a fatstock auction by a single owner.

Pedigree Cattle Sales

Pedigree breeders of both beef and dairy cattle derive a considerable part of their income by selling breeding stock. It is normal for sales to be made privately between breeders for one or two animals and, occasionally, whole herds change owners this way. However, the auction ring has an attraction because of the spectacular prices achieved if two or more bidders want to buy the same animal. As long ago as 1918, a Holstein Friesian bull bred in Canada sold there for a record figure of $106,000!

If a breeder is thinking of disposing of just a few animals, then he will probably enter them for a sale sponsored by his Breed Society. An official catalogue is prepared by the auctioneers which shows the pedigree details of the animals, their show successes, the high prices fetched by relatives at other sales and certain production details such as milk yields. When a whole herd is being sold by auction, then a sale is usually organised on the breeder's farm and the catalogue will consist entirely of his animals. This is called a Dispersal Sale and creates a good deal of interest amongst fellow breeders, as they can see in the auction ring the results of years of planned breeding. The accepted practise is that all animals must be offered for sale, unless they are sick or some other good reason is publically announced. Therefore, buyers get a chance to bid for all the very best animals in the herd. Normally, dispersal sales occur when breeders wish to retire, change breed, or possibly turn their attention to some other aspect of farming.

An alternative to the complete dispersal sale is the Reduction Sale, which can be used to raise money quickly. This is a partial dispersal where only a portion of the herd is sold. Sometimes, in the case of a dairy herd, this might be the milking portion, or alternatively, all the young stock. When the famous Mr. Charles Colling sold 47 of his Improved Shorthorn Cattle on 11th October, 1810 he made history. The sale realised £7,115.85, and the senior herd sire, 'Comet' made £1,050. From that day, record prices have continued to be broken at auction sales. This often happens when groups of breeders form syndicates to purchase bulls for use in their herds. Artificial Insemination organisations sometimes enter the bidding for bulls to join their studs.

Commercial Cattle Sales

Commercial herds can either be non-pedigree dairy or beef. The dairy herds normally try to keep as many animals milking as the farm can conveniently manage, in order to maximise milk production. Owners of these herds sometimes sell surplus groups of heifers, but more often they sell individual animals that do not conform to their standards. These 'cull' cows as they are called, are sold because they are poor milk producers, bad tempered, too old to milk economically, have weak udders or are barren. Most would be sold for slaughter and end up as cow beef, but some are sold at auction to other farmers who cannot spot the faults or are prepared to put up with them.

Poorer milk producing cows tend to be mated to beef bulls, so that their female offspring are not introduced to the milking herd. In this way, low yielding strains are not perpetuated. Also, many dairy heifers are put to beef bulls for convenience, ease of calving and because they are unknown quantities until they have milked for a lactation. As a result, the commercial dairy farmer produces a lot of calves that he cannot rear as replacements for his milking herd. These unwanted calves are sold to rearers of beef cattle, and both crossbred calves by beef bulls out of dairy cows, and certain purebred dairy bull calves, such as Friesian type, make good meat animals.

Finished beef animals that are ready for sale can be sold in a number of different ways. Butchers, abbatoirs and owners of chains of food shops all buy privately, but are sometimes prepared to sign contracts for an agreed number of animals. Premiums are paid on quality carcases that yield more saleable meat, particularly if it comes from the rump.

Alternatively, there are the auction rings where animals are sold on a combination of their liveweight, which is announced as they enter the ring, and their looks. If too many animals are entered for the same auction, then the price drops; if entries are scarce, then competitive bidding pushes the price up. Auctions are just as much a gamble for the buyer as the seller, because a good looking animal in the sale ring can turn out to be an inferior carcase, and vice versa. The fairest way would be to sell all beef animals 'on the hook' as a carcase, rather than 'on the hoof' as a lively animal in the auction ring.

BIBLIOGRAPHY

A History of Domesticated Animals, F. E. Zeuner. Hutchinson, London, 1963.

Man and Cattle (Proceedings of a Symposium on Domestication), Mourant and F. E. Zeuner, eds. Royal Anthropological Institute, London, 1963.

Adaptions of Domestic Animals, E. S. E. Hafez. Lea and Febiger, Philadelphia, 1968.

The Bull, Alan Fraser. Osprey Publishing Ltd, London, 1972.

A Dictionary of Livestock Breeds, I. L. Mason. Commonwealth Agricultural Bureaux, Farnham Royal 1969.

Farm Stock of Old, Sir Walter Gilbey, Bart. Spur Publications Ltd, London, 1976.

Modern Breeds of Livestock, H. M. Briggs. Macmillan, New York, 1958.

The Ancient White Cattle of Britain and Their Descendants, G. K. Whitehead. Faber and Faber, London, 1953.

The Cattle of Britain, Frank H. Garner. Longmans, Green and Co. Ltd, London, 1948.

European Breeds of Cattle, Vols I and II, M. H. French, I. Johansson, N. R. Joshi, and E. A. McLaughlin. F.A.O., Rome, 1966.

Types and Breeds of African Cattle, N. R. Joshi, E. A. McLaughlin and R. W. Phillips. F.A.O., Rome, 1957.

The Classification of West African Livestock, I. L. Mason. Commonwealth Agricultural Bureaux, Farnham Royal, 1951.

Native and Adapted Cattle, R. B. Kelly. Angus and Robertson, London, 1959.

The Cattle of India, Ralph W. Phillips. Journal of Heredity, Washington, 1944.

Monograph on Thari Cattle, Abdul Wahid. Dept. of Physiology, University of Karachi, 1971.

The King Ranch, Tom Lea. Little, Brown and Company, Toronto, 1957.

The Development of the Jamaica Hope Breed, Jamaica Ministry of Agriculture. Division of Livestock Research, 1972.

Dairy Cattle Breeds, Raymond B. Becker. University of Florida Press, 1973.

A Dictionary of Dairying, J. G. Davis. Leonard Hill (Books) Ltd, London, 1963.

Byproducts from Milk, Whittier and Webb. Reinhold Publishing Corporation, New York, 1950.

Milk and Milk Products, Eckles, Combs and Macy. McGraw-Hill Book Co. Inc., New York, London, 1936.

Beef Breeding, Production and Marketing, W. E. Bowden. Land Books, London, 1962.

Report on the Marketing of Cattle and Beef in England and Wales, Ministry of Agriculture, Fisheries and Food. H.M.S.O., London, 1929.

Agriculture (Eleventh Edition), J. A. S. Watson and J. A. More. Oliver and Boyd, Edinburgh, 1962.

INDEX

Numbers in bold refer to the text descriptions of the main breeds; numbers in *italic* refer to illustrations.

Aberdeen Angus *45*, **47**, 48, 49, 53, 54, 75, 149, 150, 155
Acidophilus milk 167
Advanced Register 39
African Buffalo 11, 14, **29**
Africander **122**, **123**, *127*
Ala Tau **109**, **112**, *115*
Albese 98
Alveolus 120, 121
Amrit Mahal 138
Andalusian *118*, **119**
Angeln *82*, **88**, **89**, 100, 108
Anglesey 52
Angus doddies 47
Ankole 16, **126**, *129*, *131*, **131**
Arnee **30**, **35**
Anoa, *see* Buffalo Dwarf
Artificial insemination 170, 175, 178
Artiodactyla 10
Atpadi Mahal 138
Aubrac 16, *69*, **71**, **72**
Auction 192, 193
Auer 12
Auroch 12, 13, 14, 40, 122
Ayrshire **54**, *57*, **59**

Bahima **126**, *131*
Baladi 136
Banteng 11, *34*, **35**, **36**
Bapedi **125**, *128*
Barotse *131*
Barrosa **113**, **114**, *117*
Bashi **126**, **131**
Beef 74, 75
Beef Shorthorn 59, 76, 156
Beefalo 29, **157**, *165*
Belgian Blue, *see* Central and Upper Belgian
Belted Galloway 42
Bison, American 11, 14, **28**, **29**, *31*
European 11, 14, **28**, *31*
Indian **35**
Blonde d'Aquitaine 16, **76**, **77**, *79*, *187*
Bone meal 176
Boran **126**, *129*
Bos banteng (Banteng) 11, *34*, **35**, **36**
B. bison (American Bison) 11, 14, **28**, **29**, *31*, 157
B. bonasus (European Bison) 11, 14, **28**, *31*
B. bubalus (Indian Buffalo) 11, **30**, *33*, **35**
B. gaurus (Gaur) 11, *34*, **35**
B. grunniens (Yak) 11, 14, *33*, **36**, **37**
B. indicus 10, 11, 16, 52, 122, 138
B. longifrons 16
B. namadicus 10
B. primigenius 10, 12, 16, 17
B. taurus 11
Bovidae 10
Brahman 18, 137, 138, **145**, **148**, 149, 150, *153*
Brangus **149**, *154*
Braunvieh, *see* German Brown
Brava, *see* Fighting Bull
Breed Society 138
Breikenburg 85
British Friesian 62, *63*, **64**, **65**
British Polled Hereford 48
British White 40
Brown Atlas 16
Brown Mountain Cattle 83, 84
Brown Swiss 16, **78**, *80*, **83**, **84**, 89, 109, 112, 114
Bruna Alpina 83
Brune des Alpes 83
Bubalus depressicornis (Dwarf Buffalo) 11, **29**, **30**, *32*
Buchan humlies 47
Buduma, *see* Kuri
Buffalo 10, 11, 14, 16
African 11, 14, **29**
Cape **29**, *32*
Congo **29**
Dwarf 11, **29**, **30**, *32*
Indian Water 11, 14, **30**, *33*, **35**
Butter 158, 159, 162
Buttermilk 167

Calvana 95
Canadian Holstein 64
Casein 167
Castlemartin 52
Central and Upper Belgian **100**, *103*
Champion 191
Charolais 17, 18, *70*, **72**, **73**, **76**, 100, *141*, 157
Cheese 158, 159, 167
Chianina 40, 75, *92*, **95**, **96**, 97
Chillingham 40
Churn 134

Commercial 39
Conformation 24
Controlled breeding 183
Cream 159, 162
Cultured milk 162
Cunningham 54
Czechoslovakian Red Spotted 84

Damietta 136
Danish Red **100, 101**, *103*, 108
Demonte 98
Devon 17, **53**, **54**, *56*
Dexter **61**, **62**, *166*
Dishley 40
Drakensberger **123**, **124**, *127*
Draught animals 86
Droughtmaster 18, **150**, *163*
Dunlop 54
Dutch Friesian **62**, **64**, *67*, 99, 100

East Finnish 107
East Friesland 85
Egg transplants 184
Egyptian 16, **136**, *139*

Fighting Bull *118*, **119**, **122**
Finncattle *105*, **107**, 108
Finnish Ayrshire *106*, **107**, **108**
Fleckvieh, *see* Simmental
Friesian **62**, *63*, **64**, **65**, *67*, 99, 108, 109, 112
Froment du Léon 61
Frozen semen 175

Galega, *see* Galician Blond
Galician Blond 16, **114**, *117*, **119**
Galloway **42**, *45*, **47**, 48, 54
Garonne 76
Gasconne 17, **78**
Gaur 11, *34*, **35**, 36
Gayal **35**
Geest 85
Gelbvieh, *see* German Yellow
German Brown 83
German Red Pied *82*, **85**, **88**
German Yellow 49, **89**, *91*
Gir **137**, **138**, *142*, 145, 150
Glamorgan 41
Glan-Donnersberg 89
Glycerine 176
Glyceryl trinitrate 176
Greek Steppe 78
Groningen White-headed *94*, **99**, **100**, 123
Guernsey *58*, **61**
Guzerat **137**, 145, 150

Hamitic Longhorn **16**, **124**, 133
Herd Book 38
Herd Book Number 38
Hereford *46*, **47**, **48**, 157
Hide 110
Holstein-Friesian 18, 62, **64**, *67*, 156
Horn meal 177
Hungarian Pied 84
Hungarian Simmental 84
Hungarian Steppe 78

Illawarra Shorthorn 60
Indo-Brazilian 138
Insulin 177
Inyambo **126**
Isigny 61

Jamaica Hope **155**, **156**, *164*
Jersey 16, *58*, **60**, 108, 155, 156
Judging 186

Kankrej **137**, 138, *142*, 143, 145
Kasakh Whiteheaded 112
Kefir 167
Kenya Boran **126**, *129*
Kerry **61**, **62**, *166*
Khillari **138**, *151*
Kholmogor *106*, **108**, **109**
Kigezi 131
Kirgiz 109
Kostroma 109, 112
Krishna Valley 145
Kula 84
Kuri 16, *130*, **131**, **132**
Kyloe, *see* West Highland

Lahn 89
Lake Chad, *see* Kuri
Lancashire 40
Landim 124
Leather 110, 176
Leicestershire 40
Libyan 16
Limburg 89
Limousin *70*, **76**, 77, *180*
Lincoln Red 17, **53**, *56*
Lincoln Red Shorthorn 53
Longhorn **40**, **41**, *53*
  Texas 17, **144**, *152*, *153*
Luing **156**, **157**, *165*

Maine Anjou **77**, **78**, *79*
Mammary gland 120
Mancelle 77

Marchigiana *92*, **96**, **97**
Maryuti 136
Meuse-Rhine-Ijssel *94*, **99**
Mhaswad 138
Milk 120, 121, 158
Milking machine 134
Milking parlour 134, 135
Minho, *see* Galician Blond
Miranda **113**, *116*
Mongolian **112**, **113**, *116*
Murray Grey **150**, **155**, *163*
Mysore 138

Nakali 138
N'Dama 16, **133**, **136**, *139*
Neatsfoot oil 177
Nellore 145, 150
Nguni 16, **124**, **125**, *128*
Nilotic 16, *131*
Non-pedigree 39
Nordland 102
Normandy 17, **65**, 66, *68*
North Finnish 107

Offal 176
Old Gloucestershire **41**, *44*
Oxen 86, 87

Parlour, milking 135
Pedi, *see* Bapedi
Pedigree 38, 39
Perugina 95
Piemontese *93*, **98**, **99**
Pie-Rouge de l'Est 84
Pinzgauer **90**, *91*
Podolic cattle 97
Polish Simmental 84
Production register 39
Progeny test 175
Prostaglandin 183
Pyrenean Blond 76

Quercy 76, 77

Ratinha, *see* Miranda
Red and White Friesian **65**
Red Dane 109
Red Pied Friuli 84
Red Pied Swedish 101
Red Poll 48, 54
Red Sindhi **143**, **144**, *152*
Red Steppe **109**, 112, *115*
Rennet 158, 159, 167
Rhineland 85
Romanian Simmental 84

Romagnola 17, 78, *93*, 96, **97**, **98**
Røros 102
Rotbunte, *see* German Red Pied
Russian Simmental 84

Sahiwal 155, 156
Saidi 136
Salers *69*, **71**
Sales 191, 192, 193
Sanga 124, 126
Santa Gertrudis **148**, **149**, *154*
Schwyzer 78
Semen 170
Shorthorn 18, 53, 54, *57*, **59**, **60**, 65, 72, 77, 100, 108, 109, 114, 148, 150, 155, 192
Shows 185
Simmental 18, *81*, **84**, **85**, 89, 109, 112, 114, *179*
Skim milk 159, 167
Slesvig Marsh 100
South Devon *46*, **48**, **49**, 75
South Oldenburg 85
Stade 85
Superovulation 184
Sussex 17, **49**, **52**, *55*
Swazi 124
Swedish Ayrshire 101
Swedish Red and White **101**, **102**, *104*
Syncerus caffer (African Buffalo) 11, 14, **29**

Tallow 176
Tanning 110
Tarentaise 16, **66**, *68*
Tasmanian Grey 155
Teat 120
Telemark **102**, *104*
Texas Longhorn 17, **144**, *152*, *153*
Thari, *see* Tharparkar
Tharparkar **138**, **143**, *151*
Thur 12
Toro de lidia, *see* Fighting Bull
Tsine 36

Udder 120, 121
Ukrainian Grey 109
Ukrainian Red, *see* Red Steppe
Uri 12
Urus 12

Val di Chiana 95, 96
Valdarno 95

Waldeck 85
Warwickshire 40
Watusi **126**, *131*

Welsh Black 41, **52**, **53**, *55*
West Finnish 107
West Highland **42**, *44*, 54, 156
Westphalian 85
White Fulani *130*, **133**
White Thillari 138
Wild White Park Cattle 17, **40**, *43*
Wisent, *see* Bison, European

Yak 11, 14, *33*, **36**, **37**
Yakow 36
Yellow Franconian 89
Yoghurt 167
Yugoslav Pied 84

Zebu 18, 52, 138, 143, 145, 150
Zo 36
Zomo 36
Zubr, *see* Bison, European
Zulu 124